THESE TREES
TELL A STORY

THESE TREES TELL A STORY

TELL A STORY

THE ART OF READING LANDSCAPES

NOAH CHARNEY

Yale UNIVERSITY PRESS   NEW HAVEN AND LONDON

Published with assistance from the foundation established in memory of
Philip Hamilton McMillan of the Class of 1894, Yale College.

The drawings reproduced here were made by the author with graphite and
colored pencil on paper, then digitally assembled. Photos, unless otherwise
credited, were shot by the author. Photos with MassGIS credits are courtesy
of the Bureau of Geographic Information (MassGIS), Commonwealth of
Massachusetts, Executive Office of Technology and Security Services.
Maps with USGS credits are courtesy of the U.S. Geological Survey.

Yale University Press books may be purchased in quantity for educational, busi-
ness, or promotional use. For information, please e-mail sales.press@yale.edu
(U.S. office) or sales@yaleup.co.uk (U.K. office).

Set in Source Serif type by Motto Publishing Services.
Printed in Slovenia.

Library of Congress Control Number: 2022944705
ISBN 978-0-300-23089-5 (paperback : alk. paper)

A catalogue record for this book is available from the British Library.

This paper meets the requirements of ANSI/NISO Z39.48-1992
(Permanence of Paper).

10 9 8 7 6 5 4 3 2 1

Frontispiece: Map of field sites.

For
Juno Mizuno
&
Alder Uma Isa

Publication of *These Trees Tell a Story* has been made possible with the support of the following benefactors:

Association for the Study of Literature and Environment

Lydia Rogers and Burt Adelman

The Helen Clay Frick Foundation

An Anonymous Donor

Contents

Species Checklist

The following species appear in the chapter-opening images. See which you can spot before you read each chapter.

Trees
Tupelo (Black Gum), *Nyssa sylvatica*
Paper Birch, *Betula papyrifera*
American Chestnut, *Castanea dentata*
American Beech, *Fagus grandifolia*
Swamp White Oak, *Quercus bicolor*
Scarlet Oak, *Quercus coccinea*
Chestnut Oak, *Quercus montana*
Pin Oak, *Quercus palustris*
Northern Red Oak, *Quercus rubra*
White Ash, *Fraxinus americana*
Eastern Cottonwood, *Populus deltoides*
Quaking Aspen, *Populus tremuloides*
American Basswood, *Tilia americana*
White Spruce, *Picea glauca*
Red Spruce, *Picea rubens*
Pitch Pine, *Pinus rigida*
White Pine, *Pinus strobus*
Eastern Hemlock, *Tsuga canadensis*
Boxelder (Ash-Leaf Maple), *Acer negundo*
Silver Maple, *Acer saccharinum*
Sugar Maple, *Acer saccharum*

Shrubs
Oriental Bittersweet, *Celastrus orbiculatus*
Leatherleaf, *Chamaedaphne calyculata*
Mountain Laurel, *Kalmia latifolia*
Lowbush (Hillside) Blueberry, *Vaccinium pallidum*
Highbush Blueberry, *Vaccinium corymbosum*
Border Privet, *Ligustrum obtusifolium*
Multiflora Rose, *Rosa multiflora*
Poison Sumac, *Toxicodendron vernix*

Forbs
Virginia Glasswort, *Salicornia depressa*
Wild Cucumber, *Echinocystis lobata*
Purple Pitcher Plant, *Sarracenia purpurea*
Hog Peanut, *Amphicarpaea bracteata*
Royal Fern, *Osmunda spectabilis*
Tawny Cottongrass, *Eriophorum virginicum*
Common Reed, *Phragmites australis*
Broad-Leaf Cattail, *Typha latifolia*
Bracken Fern, *Pteridium aquilinum*
Round-Lobed Hepatica, *Hepatica americana*
Canadian Wood Nettle, *Laportea canadensis*

Rocks
Basalt
Granite
Quartzite
Schist
Arkosic Conglomerate

Vertebrates (including tracks/signs)
Wood Frog, *Rana sylvatica*
Osprey, *Pandion haliaetus*
Moose, *Alces alces*
White-Tailed Deer, *Odocoileus virginianus*
Eastern Coyote, *Canis latrans*
Eastern cottontail, *Sylvilagus floridanus*
Bobcat, *Lynx rufus*
Striped Skunk, *Mephitis mephitis*
Black Bear, *Ursus americanus*
Human, *Homo sapiens*

Invertebrates (including tracks/signs)
Dwarf Antmimic Spider, *Phrurotimpus* sp.
American Giant Millipede, *Narceus americanus/annularis*
Puritan Tiger Beetle, *Ellipsoptera puritana*
Oak Gall Wasp, Cynipidae
Blackgum Leafslug Sawfly, *Caliroa nyssae* (likely)
Leaf-Mining Sawfly, *Pseudodineura parva*
Paper-Making Wasp, Vespidae
Pygmy Leaf Mining Moth, *Stigmella* sp.

THESE TREES
TELL A STORY

(overleaf)
Figure 1.1. Know thy land.

1. Home

"We're thinking of cutting down our trees." Caleb's voice suggests a tinge of guilt, as if he's looking for me to talk him out of it. Comfortable playing the role of the Lorax, I take the bait.

Caleb and Maya have good reasons. The trees might fall on their house—if an ice storm, tornado, or a crushing wind microburst drops down. Part of the house is rotting out from below, and fixing the rot requires digging a drainage ditch where one of the trees stands. The yard is too shady—they want to grow apples and their orchard would need more sun. The bumpy roots protruding above ground level make it hard to mow. They don't really like pine trees anyway. They'd leave the oaks because oaks at least make acorns. But what good are pines?

I go to work defending the trees. They can regulate temperature—keeping your house cool in summer, warm in winter—by blocking sun and wind. You don't want your child getting sunburned in your own yard, do you? Cool shade is ideal for a summer playground. Growing trees suck carbon out of the air—that's where all the weight of their wood comes from—offsetting the greenhouse gas emissions that your family produces. Because the grass grows more slowly, you don't need to mow as often under trees. The wildlife depends on these trees. White Pine is an awesome native species—the Haudenosaunee named it the Tree of Peace.

Think of the flocks of winter resident birds looking for a good evergreen shelter from the cold.

Their orchard argument is hardest to rebut. Caleb grew up in an orchard. Isn't local food key to a sustainable future? Since they moved in, Caleb has been working to get a productive garden going. But it's a struggle. The clay-filled soil is so wet that he resorted to bringing in a dump truck of topsoil and heaping it up so that vegetables would grow above ground level out of the water. Doesn't he deserve his dream orchard landscape?

WILD YARD

To be fair, I should note that Maya and Caleb are very eco-conscious, with every desire to do right by the world. Besides, I'm in no position to act high and mighty here—my own yard bakes in direct sunlight. When Sydne and I moved into our house ten years ago, the front lawn was mainly a large, green wasteland of grass. Just like our neighbor's, it had been mowed regularly to keep a nice trim profile. Being dominated by a single non-native grass species, the lawn invited few interesting animals to play. Then we moved in. Instead of buying a lawnmower, we bought a "wheeled trimmer," which looks basically the same but works on meadows and raspberry thickets. Today most of our yard hasn't been cut in ten years and trees are coming back (fig. 1.2). Some parts we still trim regularly as lawn—but even there, just as my dad once did for me when he mowed, we dodge wildflower patches for our kids to study.

We put a sign out front that says "Wildlife Habitat," and it's no lie. We have two species of foxes, coyotes, bears, turkeys, hatching monarchs, and endangered snakes. Birds feast on the pollinators and other bugs that overwinter in unraked leaves and abandoned raspberry canes. On summer evenings our yard is aglow with sparkling fireflies in all corners, whereas our neighbor's yard

Figure 1.2. The border between our yard and our neighbor's.

is just dark. Our yard is brimming with songs of myriad species of crickets and grasshoppers, in contrast to the monotonic hush next door. Our kids grow their minds by exploring a maze cut through the goldenrod. The neighbor's kids drive go-carts in never ending circles through their lawn—OK, I'll admit that does look like fun. And I can feel the glare from their second-story window looking down at these irresponsible neighbors who can't take care of their mess of a yard. But, I protest, we're not lazy. We're consciously putting our yard to work. It's sequestering carbon, supporting life, and enriching ours.

It was Susannah Lerman who inspired me to put the sign in front of our yard. Every day on her commute, she would drive past our house. She and I studied in the same graduate research lab—that of urban ecologist Paige Warren. Much of the time, though, one of us would be out in the field far from the lab. I was busy chasing rare salamanders across the landscape. Susannah studied urban birds in Phoenix and ways to better connect city people with the species around them. Our labmate, Rachel Danford, spent her time surveying Boston residents to see what it would take to convince people to leave more dead branches on their trees for woodpeckers to nest in.

One day Susannah remarked to me that she loved driving past our unkempt lawn but wondered how the neighbors felt. Particularly "Lawnmower Man." He seemed to mow his lawn nearly every day, and I swear I once saw him mow it twice in one day—though perhaps that was more about riding the machine around than actually cutting grass. Or perhaps, seething in his seat, fingers clenched around his grips, he mowed for cathartic relief from the out-of-control nightmare unfolding on the other side of his property line. As Susannah had found in her own academic work, people often see wild yards as a blight—a betrayal of the neighborhood. But if you frame your yard appropriately, such as with a sign out front explaining your purpose, people often see it in a whole new light.

PHOENIX

With our graduate school days in my thoughts, I went to visit Susannah this year to hear again about her research, still stuck in my mind from many years ago. I opened the doors of the government complex where she now works and wound my way through the dim cubicles. Unsure of where her office lay, I crept uneasily around corners, feeling a bit like Ralph, the motorcycle-riding mouse who had been forced to run a maze in that week's bedtime reading to my son Juno. Rather than peanut butter, my reward at the end would be Susannah's insights into how homeowners unknowingly shape the bird universe.

Last I saw Susannah, she was running a project that involved mowing lawns in Springfield. She mowed different lawns at either one-week, two-week, or three-week intervals and then counted the numbers of flowers and bees in each lawn. Her results showed more flowers with a three-week rotation but more bees with a two-week rotation, although those two-week bees were dominated by just a few species. Some people worried about whether letting their lawn go would bring in more ticks. In that urban setting Susannah found no ticks in any yards. In our yard, unfortunately, ticks are a real concern that weighs on us daily.

I found Susannah's office at the end of a corridor, with Susannah hard at work inside. Next to her computer, a black frame on the desk held a *New Yorker* cover illustrating a man sitting on a city park bench. Above the man, a bird in a tree sang the sweet songs of spring. The man sat listening to his iPod, oblivious. Next to this image, a smaller frame on Susannah's desk was angled such that only she could see the picture, but I assumed it was a photograph of her son, Matan. As a three-month-old, Matan had accompanied Susannah for the intense fieldwork I had come to ask her about—work that began with a 2,500-mile journey to the Sonoran desert. In fact, a burning question on my mind was how Susannah had managed to pull off that work with an infant in her arms, alone.

A few days later I found myself in that same desert 2,500 miles away with my own children. We had flown to Phoenix, Arizona, to retrace Susannah's steps from years prior. For our single day there, we faced an ambitious schedule, squeezed between a 10:00 AM rental car pickup and a 3:00 PM drive to the Tucson airport. In those five hours, we needed to track down the four types of yards that drove Susannah's findings. As we talked in her office a few days prior, Susannah had pulled up Google Street View on her computer and walked me virtually through different Phoenix landscapes, pointing me to neighborhoods we should visit. Of course, in our five hours we also had to factor in time to eat at Susannah's recommended lunch spot, Señor Taco, and time to visit Susannah's recommended kid destination, the Musical Instrument Museum. Oh, and we needed to stop at REI to buy a new sun hat for Juno.

Before we even pulled out of the airport hotel parking lot, we started to see the important pattern Susannah had documented: house sparrows and starlings in the hotel's big trees. The hotel, like the typical mesic yard we were headed for later in the day, was landscaped in a manner totally inconsistent with the Sonoran desert. Mesic environments are defined by moderate amounts of water, unlike a desert. In a mesic yard you find expanses of grass under the shade of broad-leaved trees bordered by lush shrubs and flowers. Mesic homeowners seem to have a nostalgia for the eastern forest ecoregion; they use their garden hoses to transform the desert. The green and shady mesic yards we visited could just as easily have been in Baltimore. Xeric yards, by contrast, conform more to the desert. "Xeric" means "dry," and such yards are typified by desert and drought-tolerant plants (fig. 1.3). Beyond the obvious waste of water, Susannah found that the effects of mesic yards rippled throughout the ecological community.

I picture Susannah patrolling from house to house, holding Matan in one arm, a clipboard in another arm, and writing with a pencil held in the hand of her third arm as she recorded the extent to which native desert shrubs and cacti had been replaced by nonnative plants in mesic yards. Then, waiting for the moments when Matan was quiet, Susannah would listen for the birds and record the species and abundances of all she heard and saw. Sometimes, when Matan was asleep in his car seat, Susannah would conduct surveys through her open window, watching the experimental feeding trays she'd set up to understand the underlying rules controlling which birds choose which yards. As she delicately conducted her scientific tests of ecological theory, she had to always hope that the infant wouldn't wake up screaming and send the birds scattering.

Not wanting to wake the baby is why Susannah didn't honk to prevent the all-too-preventable collision that marked, as Susannah sarcastically put it, the "high point" of her fieldwork. Surveying the birds from inside her parked car, Susannah watched in slow motion as a man slowly, very slowly, backed his car out of his driveway, across the street, and directly into her car. He then drove off without stopping. That day, like every other day that Susannah was alone in the desert caring for her infant while establishing her scientific career, she cried.

Susannah's commitment to her work paid off. Her data confirmed the stories people on the streets told. In the mesic neighborhoods people complained to Susannah about the noisy, messy flocks of pigeons, starlings, and house sparrows (fig. 1.4). Hearing that she was an ornithologist, they asked Susannah how to get rid of these birds. The mesic neighborhoods were full of nonnative species, and the residents detested their birds. But in the xeric neighborhoods people delighted at the roadrunners, cactus wrens, and Gambel's quail. These xeric neighborhoods were full of desert-adapted native species, and the residents loved their birds (fig. 1.5). Native plants support native wildlife. Cactus wrens and gila woodpeckers, for instance, are adapted to nest in cholla cacti and saguaro cacti. If you lose the special plants, you lose the special bugs, mammals, and birds. And people care.

But there are many layers to this story. Even when Phoenix homeowners plant desert species, they don't necessarily use less water. As my family and I drove through one of the xeric neighborhoods, a roadrunner darted down the sidewalk and underneath a spherically pruned green desert shrub. The desert birds were here, but this sure didn't look like a wild desert. The homeowners had carefully laid down uniform gravel instead of grass, placed cacti around meticulously pruned desert trees—although the cacti and tree species may have been borrowed from deserts on other continents—and maintained exhaustive control of the environment. In such yards homeowners often add water all year round to keep the vegetation green, negating the water-saving advantages of xeric yards. But desert plants are designed to go dormant for long stretches of the year; they don't need hoses and sprinklers.

Figure 1.3. Phoenix yard types: (a) wild xeric, (b) landscaped xeric, (c) barren, and (d) mesic.

In the wealthy Scottsdale neighborhood, I ducked through a fence of a gated community as a flock of Gambel's quail bobbed their heads under untamed, browning desert hackberry in someone's yard. This yard demonstrated the wilder approach to lawn care, where nature does all the work. Plants are allowed to go

brown like the surrounding desert in the dry season, and it's these yards that both support native species and save water.

I needed to capture a picture of the whole lot in context, so I headed back out to the sidewalk. There, I stood up on my tiptoes and held my camera over the massive wall that kept intruders out.

Figure 1.4. (a) Pigeon, and (b) house sparrow, exotic species in North America.

But my over-loved camera lens malfunctioned. I removed the lens from the body, smacked it against my leg, and shook it until the loose piece inside rattled to a new position—a trick that usually makes it work temporarily. The maneuver failing, I went back to the car and asked three-year-old Juno if I could borrow his camera.

All I wanted to do was take the picture and get out of there without finding myself in trouble, again. As a disheveled dude peering over an affluent fortress wall taking pictures of someone's estate, I didn't want to attract any more attention than necessary. I recalled Susannah's story of cops being called on her doing this fieldwork. Of course, the baby strapped to her front softened her image—a strategy I employ as well when I can.

As a field ecologist, I've had my own share of run-ins. There was the time Charley Eiseman and I were scheduled to present an evening slideshow on bug tracks at a church in the dead of winter. We showed up a couple hours early, as we always do, taking pictures of the miniature poops, cocoons, webs, and eggs on the outside wall of the church to incorporate into our slideshow. As the two of us crept slowly around the perimeter of the building staring at tiny bugs in the dimming light, an elderly woman working inside called 911. But just after she called, we left for a short pizza break, oblivious to our escalating peril. An hour later we were back tak-

Figure 1.5. (a) Roadrunner, and (b) curvebilled thrasher, native species of the southwestern United States.

ing pictures of bug tracks on the church wall when a pair of policemen, having received a second 911 call from a terrified woman hiding under a church desk, came skidding up in their car and jumped out to apprehend us while their wheels were still rolling.

Or there was that time Sydne and I parked our rental pickup along a remote edge of forest above the river separating the United States and Canada. As we were hunkering at the water's edge sampling federally endangered plants, the border patrol came creeping up on us through the woods. We started yelling at them not to step on the plants, but they didn't seem to understand.

Oh, and what about that time in rural Vermont when Charley and I, parked at the end of a one-house road at 5:00 AM to begin the hike in to our assigned survey plot, were chased off by a stern woman from the house wielding a large silver revolver. Or that time I was strangled in the hills of Mendocino, California, by the two wiry hands of a long-haired man defending his property while the German shepherd I was hiking with wagged her tail next to me. I could go on.

But here's my broader point: as an ecologist, you can't forget about society.

It took us more than the allotted five hours to wrap up our Phoenix scavenger hunt, but we did succeed in witnessing the four yard types: wild xeric, landscaped xeric, barren, and mesic. We then raced down to Tucson to pick up my mom from the airport so that she could help watch the kids the following day while I headed off to the Mexican border.

The next day, with the long metal border fence snaking over low hills into the distance, Chelsea Mahnk held a metal antenna high in the air and listened through static on her transceiver for imperceptible beeps. Occasionally, border patrol SUVs drove by dragging giant tires along the dusty road to create a clean surface for finding footprints of migrating humans. But Chelsea and I were after a different migrating species.

Simply creating the right habitat for wildlife in your yard isn't always enough. The animals also need to be able to get there. For winged creatures maybe it's not so hard to fly over asphalt wastelands and plop down in a little glowing paradise. But down at the border fence, we were tracking turtles. Arizona mud turtles. These turtles migrate seasonally between desert ponds and uplands where they hole up in kangaroo rat burrows. The thing is, some of their upland habitat is over in Mexico. The fence in that spot is still open enough that the turtles can slip through—although we can't. The turtles were carrying radio transmitters, and we were straining to triangulate a position on them in Mexico from where we stood north of the border in the Buenos Aires National Wildlife Refuge. Later we planned to look down with Google Earth and try to understand their habitat needs. What, we wondered, will happen to the turtle populations if this fence becomes a less porous wall?

CONNECTING NATURE AND PEOPLE

As night set in, Chelsea and I headed back to Tucson, where I rejoined my family for a vacation in the desert. One of the rituals during this vacation was to get up at sunrise and head over to the bird ramada—a covered viewing platform surrounded by bird feeders. The feeders attracted many special desert birds, like quail, cactus wren, and curve-billed thrashers, and some other birds more familiar to me, like house finches.

Bird feeders. As a citizen, bird feeders are a great way to support wild animals and connect with nature. As a scientist, however, I see bird feeders as a great experiment. What species are we really supporting? Do the increased numbers of commercial sunflower seed-eating species, like house finches, crowd out other species, like phainopeplas, that specialize in wild bugs and desert fruits? Bird feeders have certainly helped expand the range of house finches from their native West to a continent-wide takeover. And what are we doing to the species as we feed them? Even as house finches rely on eastern bird feeders, researchers have shown that bird feeders facilitate the spread of a deadly eye disease, conjunctivitis, among the finches. Like a snot-covered daycare room, feeders sometimes give all the visiting finches pink-eye.

More than just altering which species live where, with bird feeders we are altering the very species themselves. Researchers in Phoenix found that the house finches within the city were evolving larger, stronger beaks to crack through thick sunflower seeds, compared to the house finches still fending for themselves in the wild desert. With a changing bill size, the urban finches are now evolving different songs. At the same time other urban forces are causing city finches to grow less colorful feathers than their country cousins.

And then there are the social science questions about bird feeders. That is, what parts of our society participate in the feeding of birds? Who gets to enjoy which birds? In Phoenix the social justice component of ecology is impossible to ignore. The richest neighborhoods have the best native birds. Along the mesic-to-xeric gradient, there are some advantages to yards with moisture-loving eastern trees—most notably that they keep your house a bit cooler in the hot desert. However, the poorer Phoenix neighborhoods don't even have that.

As my family and I drove through the poorest neighborhoods, we saw vast expanses of bare dirt, pavement, and gravel. No plants, no interesting desert birds, no cooling non-native trees, but lots of pigeon poop. The poorest peoples living in the most degraded environments. Beyond the desert, this is a pattern that holds true across many contexts.

BOSTON

In mid-May I pay a visit to my aunt and uncle in Boston. Loretta is an atmospheric chemist and Michael is a retired doctor but perpetual activist who has lived in this one tiny apartment for decades. He is a model citizen for the environment—and at times a bit fanatical about it. As I was growing up, I remember Michael pulling trash out of garbage cans to reuse, taking public transportation at all costs, riding his bike whenever he could, and minimizing his footprint by packing himself in among all the little bricks in the city.

Here's the paradox that Michael and Loretta's lifestyle presents for me. If you love nature, the best thing you can do for the environment is to live in a city. But, living in a city, we lose our connection to nature. We don't get to reap all its benefits, and I worry that we lose our ability to care for it appropriately. I've lived for brief periods of time in urban settings. At first it's sort of fun. But then I go crazy.

When I arrive at the Boston apartment, Lorretta is away in her office, presumably working out the fates of chemicals floating around in the sky. Michael is home, and he takes me out for a walk around the neighborhood. Knowing of my need for nature, he leads me to a small park where a few dozen of Boston's trees stand above a cleared understory. Red oaks, beeches, and some sort of pine with deeply furrowed bark. They seem to be in the hundred-year-old range but have an odd form. Most start out single-trunked like normal forest-grown trees. Then, about ten feet up, they split into multiple trunks. We scratch our heads trying to reconstruct the history of the land.

From across the park I notice a telltale discolored stripe running up the side of the pine tree. I point it out to Michael and bring him over to show how it's made of hundreds of little nicks, each half the width of a pencil. Each of these, in turn, consists of a pair of two grooves. The nicks are positioned at all sorts of angles and come in many shades of brown—from a rich mahogany to a weathered gray.

In all the years that Charley and I have taught animal tracking, we must have brought hundreds of students to trees just like this. As they stare at the marks on the ash tree next to one of our favorite spots—where coyotes once barfed up a pile of voles—we ask the students probing questions about what they notice, hoping someone will figure it out. Nobody ever knows what this is, even though they've all walked by such trees dozens of times, even though it's a blaringly obvious sign, and even though they're all very familiar with the animal that created it.

Of course, Charley and I didn't know this sign either until tracker John McCarter pointed it out to us. It's the territorial marking sign of gray squirrels. Generations of squirrels over many decades have come to this pine, sniffed, taken a quick bite with their incisors, then rubbed their cheeks into the bark to squeeze out pheromones from their scent glands. What does it all mean? Nobody knows. John told us that when he first discovered this, he broke off a piece of bark and brought it home. The squirrels in his yard went crazy over the scents on that bark. I've tried without success to recreate that experiment.

Michael leads me over to a depression that he says fills with water sometimes. The ground is a bit soft and covered with lesser celandine, an invasive spring ephemeral that takes over wetland margins. But there's not enough water standing now to support much life. Before all the roads around us diverted the rain into gutters, would spotted salamanders have bred here in a healthy vernal pool? Probably. Today there are no spotted salamanders left anywhere within dozens of miles of us.

Michael wants to show me the nearby greenway, so we start hiking north. We climb up and down a series of steep hills and wonder what these are. Michael suggests that they are drumlins—glacial formations created beneath a moving ice sheet. I'm tired, worried about my knees, and find it hard to keep up with Michael, who has over thirty years on me. I blame it on lack of sleep and the physical demands of having two little kids.

We get to the greenway—a paved bike path edged with two narrow strips of green on either side. Michael says that this is what passes for nature in the city. "Invasive-dominated degraded ecosystem," I say. We walk for a while, then come to an area where the ground alongside the pavement is muddy and the pedestrians are shaded by silver maples, red maples, elms, and a cottonwood. Floodplain trees. I close my eyes and see this spot as it once was: a forested wetland full of chattering wood frogs near a river. I open my eyes and see joggers, baby strollers, dogs on leashes, and asphalt painted down the middle with a solid yellow line.

HOME AGAIN

Back where we started, Caleb and Maya are still contemplating dusting off the chainsaws (see fig. 1.1). Despite my tree-hugging impulses, as a conservationist I can't be against cutting trees. After all, I rely on a woodstove to heat my house. Wood is a local, renewable resource. When harvested sustainably, burning wood can be a much more carbon-neutral way to heat your home than burning fossil fuel. The carbon I put into the air through my woodstove is recaptured by new trees growing out on the landscape. Modern conservation sees humans as a critical, interconnected part of protected landscapes. The goal is not to preserve nature untouched but to build a strong relationship between people and nature, making both healthier. And it all starts at home. In the United States more than 80 percent of people live in cities and suburbs, so it's in yards and urban green spaces that we must start. Home is where each individual can have the greatest control over our planet.

With this in mind, I know now what I want to say.

Dear Maya and Caleb,

Here's what it comes down to for me: if you're going to cut down your trees, you should at least get to know them first. More than just learning all the good things trees do for us, I want you to know these particular trees in the context of the whole system. How did these individual trees get here? Why these species, and where do they fit in a broader ecological narrative? How do wild animals use them?

And while you're at it, I want you to know more than the trees. How are the trees controlled by and influencing the rest of the parts of your yard? Why is your soil so wet? Where did the soil even come from? Where does the water that's rotting your house come from? Where is the water going?

And I want you to know more than just your yard. How do the features of your yard relate to your neighbors', to the community, and to the whole region?

And I want you to know your yard in time. How was your yard shaped by events in the past fifty years? Hundred years? Thousand years? Million years? Billion years? Where is your yard going? What will it look like in fifty, one hundred, one thousand years? I want you to see your yard as a bridge across space and time.

The pines, the oaks, the mud, the water, the land. It's not random but all part of a long, unfolding story that you have a role in. Dig up the details.

Then, only then, will I trust you as the shepherd of your yard. In managing your yard you control a whole universe, and I want you to know that potential. Like giving thanks before killing and eating a deer. Like saying a blessing over your bread. Like actually stopping to absorb the artwork on the wall of the museum you've paid so much to enter. Know thy land.

Love,
Noah

Knowing your own land. That's what this book is about. It's not really about facts, names, or specific places. It's about how to think about a landscape, what questions to ask, and how to connect the dots. And it's about your place in the landscape. This chapter isn't even about Caleb's yard, it's about your yard. Whether you own a house, rent an apartment, carry a tent, or sleep in a shelter, your land has stories to tell. Yard, sidewalk, park, mountain, or ocean, can you read the stories?

THIS BOOK

The best job I ever had was teaching a college course, *Field Naturalist*. Each Friday we would drive the van to a different site in the Connecticut River Valley where the students would be confronted with a mystery to solve. At each of these sites, we discovered the underlying stories of ancient volcanoes, glacial lakes, farming, logging, and other forces. Meanwhile the students' semester-long project was to tell the story of their own home site on campus, returning from our class field trips with new layers of questions each week.

This book is that course. Each chapter is a trip to one of the class field sites. The mystery of each site is depicted in the chapters' opening images, complete with the clues the students used to solve the puzzle. The home project site is yours to choose.

See what you can make of the opening images, imperfect as they are, before you read the chapters. What pieces do you see? Can you find any visual patterns, shapes, or gradients that might tell a story? Can you identify any species? What processes might be at play in forming the patterns? How might you value this landscape? I'll drop hints throughout the writing—see if you can pick them up to solve the puzzles or at least make some guesses. Then use the prompts at the end of each chapter to reflect on your own home site.

For the rest of this book, we'll be primarily exploring wild landscapes. But, just as in Boston and Phoenix, even if your world is paved over and penned in, you can learn to read the urban landscape with much the same approach.

If you're new to nature, don't worry about the particulars of species names and unfamiliar concepts. Just look for the visual patterns and try to imagine what the site might feel like. If you're a seasoned naturalist, go ahead and break out the field guides and try to nail down the species and habitat preferences—for a leg up, start with the plants in the table at the beginning of this book, which contains all the species you'll encounter in the opening images. If you're like me and only have time for audiobooks, well, just enjoy the stories, although the images are also posted on this book's website along with many more that couldn't fit in the printed book. If you're an educator, check out the appendices on the website and don't overlook the references—I've tried to document every idea from human-chasing moths to butt-itching roses.

Everything in this book is real. I promise not to cheat—although I do love fair tricks. These are the actual field sites from my course, the images were all taken on location, and everything is told just as it happened. These aren't stylized versions of nature; these are the thing itself, with many different layers overlapping and interacting. Likewise, the interpretations I give are my best shot at it and could be wrong. What I'm after here is not the answer but the process of critically engaging a landscape with all the perils that brings.

Wandering into my special places in nature, I find a sort of timelessness. This slips into our language when Charley and I talk about such places. We typically use the present tense in our conversations about whatever we, our students, or the other creatures did in a place, no matter how long ago we witnessed it. The memories always linger in the present. Among other things, it is a mirror to the structure of nature—if something happened once at a place, it's likely to happen again and again there. You'll notice that I've woven a small bit of our linguistic convention into the structure of this book in order to bring the field sites to the foreground and to help you experience their timelessness.

In college the most important thing I did was to join an informal band of students and locals dedicated to primitive skills such as animal tracking, wilderness survival, and generally knowing nature in a deep and intimate way. The Woodsy Club. At the start of any meeting or walk outside, we would express some creative form of the "Thanksgiving Address," an idea shared with us from Haudenosaunee traditions. Beyond anything magical, I see it as a little meditation—a way to slow down before starting something important to ensure it's done carefully and with focus. It's the same basic idea behind reciting a blessing before your meal, at least in my understanding of Jewish mysticism. But in my Sunday school, they never taught me the Hebrew blessing to say before wandering in the woods. The Thanksgiving Address filled that void for me.

In the Woodsy Club we didn't always stick to a specific script, but the thanksgiving would often follow the basic pattern of moving from bottom to top through the layers. We would start by giving thanks to the dirt and rocks, then acknowledge the plants and trees, then the little animals and big animals, the people, the birds, the weather, the sun, moon, and stars, the great mysteries, unknown forces, and a blanket thanks for anything else we forgot. For the basic skills that we practiced in the Woodsy Club, it felt essential. For whatever reason, I don't think I've ever successfully made fire-by-friction on occasions when I didn't take the time to center myself with a little thanksgiving first. And I've found this really prac-

tical as an ecologist—it reminds me to think about all the layers of interaction. Just before the path leads you into the woods, take a moment to pause, step back, and look at the whole picture (fig. 1.6).

Figure 1.6. *Field Naturalist* students on the last day of class.

MAJOR LESSONS FOR INTERPRETING A LANDSCAPE

- Think about how you manage your own yard and the ripple effects it has through the whole ecosystem.
- Even where nature seems paved over or mowed down, look for clues—like wetland trees still hanging on—that might tell a deeper story.

Figure 2.1. A puzzle.

2. Land

LOSING THE TRAIL

It's an early August morning. The sun hasn't yet risen, but the birds have begun their dawn chorus. A few miles south of Caleb and Maya's house, I'm hiking up a forested trail carrying Juno on my back, already feeling the heat of the day. As the trail turns up a steep slope and the sweat starts to seep from my forehead, we abandon the path to take a less steep route off-trail along the landscape contours.

Following the trail is the easiest way to be lost. Sure, that trail might take us to a preordained destination faster, but we'll have no idea where we are when we get there. While we're on the trail, we lose track of what's around us and where we are in space—we are lost. We put our trust in the trail, ceding responsibility. We give up our awareness, our senses, our minds. Our interface with the landscape boils down to just two numbers: the total length of the trail and the distance we've traveled. Staring at the path a few feet in front of us, we are not fully engaged with the surrounding world.

Step off that path and suddenly we have to look up. Look at the shape of the land and decide how steeply we want to climb. Look at the trees in the distance and pick a target to walk toward. Keep looking behind so that we will recognize the forest when we encounter it from the other direction on our return trip. Study the shrub layer for gaps to duck through, following the occasional animal trails worn through the denser areas. Use the network of deer paths when traversing steep slopes to gain level footing. Keep an eye out for poison ivy, rose thorns, and ticks waving their arms in hopes of catching a ride. Study the patterns of light for clearings. Monitor the changing habitats near and far: white tops of sycamores in the distance signaling a creek; chestnut oaks nearby telling us we've reached the drier hilltops; the banjo-like plunk of a lone green frog calling from the wetland ahead that we hope to steer around. Keep an eye on the rising sun and remember where south is as we walk. This whole time, we maintain a map of the landscape in our heads, filling in the details as we go. That is how we get to know the world and our place in it.

OPPOSING SLOPES

This morning we hug the edge of a small valley, skirting a wet seep a few feet below. We gradually ascend as the valley narrows until it can narrow no more—we are standing in the bottom of a small "V." I raise my camera to my eye and capture the center photograph for the chapter-opening image (fig. 2.1). I think I'll call this spot Bark Hollow.

There's a mystery here. Looking on my camera LCD, I can see that I've captured it. It would be easy enough to miss, even in this photo. If you're not paying attention to the landscape, if you're not tuned in to vegetation, if you're not looking for patterns, if you don't stop. But once you know to look, it's impossible not to see the pattern in the image. It's not a product of the lighting; the pattern is real. The vegetation on the right is entirely different from the vegetation on the left. Here, in my *Field Naturalist* course, this left-right pattern was the central puzzle. We went home only when the students' mounting observations finally produced a narrative that coherently described the pattern.

The first thing I always ask my students to do when we get outside is to orient themselves. Playing in my head is that R.E.M. song "Stand," which asks us to face north, consider direction, question why we haven't before, and then look to the sun for help. But when I ask my students to check the sun for directions, they laugh as if this is just another of my sly jokes—because this must be an exceptionally difficult feat, right? Our lives are directed by walls, clocks, and GPS navigation. The sun has become meaningless. But doesn't the sun still rule our lives, even if we never look to it? Why wasn't it until college that I first understood the simple path of the sun through the sky? And it wasn't even college proper—it was that strange, unofficial student group, the Woodsy Club, that taught me to see the sun.

Where is the sun at noon? Most folks know that the sun rises in the east, generally, and sets in the west, generally. Where does it go for the rest of the day? If we are north of the Tropic of Cancer, which cuts through the center of Mexico, then the sun will be due south and our shadow will point north at noon—at least if it's true solar noon, defined by when the sun is highest in the sky and not by geopolitical time zones. Up at these northern latitudes, the sun spends the entire day in the southern part of the sky. And if you face the sun, it always moves from left to right. But not when you're in the Southern Hemisphere.

For our honeymoon Sydne and I took a trip around the world— funded by $10,000 we won with a spur-of-the-moment scratch ticket. We headed first through South America. Down there the landscape, species, and societies were all a bit different from those at home but not really that different. Squinting our eyes a bit, we could convince ourselves that we were still just in some far-off town in the United States. That is, until we looked up.

The real immutable difference is in the sky. At night it's not just that the constellations are arranged differently in the Southern Hemisphere, but the objects are of an entirely different quality. For the first time we saw the other half of the universe, and it was beautiful. Big, blurry galaxies hovered over us like clouds that wouldn't go away. Why don't we have anything that cool in the northern half of the universe? We were staring directly at the bright center of the Milky Way, partly obscured by dark nebulae hanging in front of it. A few familiar constellations, like Orion, still danced upside down across the northern part of the sky. But as we hiked along alpine ridges in the Andes, the dippers were nowhere to be found. And the South Star was missing. Whereas everything in the northern sky spins around Polaris (the North Star), down south, that center point was blank (although in a few thousand years, Earth's wobble will have reversed these fortunes). Most fundamentally, there was no denying that the stars turned the wrong way and the sun was moving backward. It was deeply disorienting and, on some level, horrifying.

ASPECT

In the Northern Hemisphere, if you're standing outside facing the sun at noon, you'll be facing south. Close your eyes and feel the heat on your face.

The sun's rays are, of course, what feeds plants. Each plant twists and stretches and fights its neighbors to get its little piece of the sun's light. But, as we know, too much sun isn't always good. The sun's heat causes water to evaporate, which can quickly turn a happy, lush plant into a crisp. Evaporation, it turns out, is one of the greatest threats to plants. You may be familiar with special adaptations that cacti and other succulent desert plants have, such as thick, waxy, bristly leaves to store water and reduce wind-driven evaporation. But even in lush forests, nonsucculent plants have tricks to hold on to water vapor, like folding their leaves and closing their pores at night when they need to breathe in less CO_2. Water retention is the reason evergreens of the north have round, waxy needles. It's a reason that many alpine plants are formed in miniature, round pincushions. It's a primary reason that deciduous trees shed their leaves in winter. Increased evaporation is why we expect many of the world's forests to grow more slowly as the globe warms. In short, water availability is a huge driver of ecological patterns.

Within any broad climate the local microclimate depends on many factors, such as the amount of sun, wind, and water that a site is exposed to. Up on the hill behind our house, there is a five-foot-wide and thirty-foot-deep rock crevice where the sun seldom shines. In that crevice the snow lasts months longer than the snow everywhere else around, and a whole set of mosses and ferns grow there that don't grow on the landscape nearby. To a lesser extent every hill, rock, hole, field, forest, building, and road influences the microclimate of the plants growing on or next to it.

While many species occur over broad geographic ranges, they often can only be found in specific habitats within that region. Suites of species adapted to similar habitats usually occur together. We talk about such "natural communities" and the key indicator species for any given habitat. Open any field guide and, in addition to telling you how to identify species, it will usually tell you the particular habitat type that the species belong in. Using two such guides, I looked up the site preferences for each of the species depicted in our chapter-opening image and put the results into figure 2.2. See for yourself.

One fundamental microclimate difference is between slopes that face north and slopes that face south. Just as the sun warms your cheeks as you face it, the sun warms and dries the soil of a south-facing slope. North-facing slopes, like the back of your neck,

White Ash
Rich,
Moist
Sites

Basswood
Rich, Moist
Sites

Hog Peanut
Mesic
Forests

Round-Lobed Hepatica
Rich, Mesic
Sites

American
Chestnut
Warm, Dry
Sites

Chestnut Oak
Hot, Dry
Sites

Mountain Laurel
Acidic, Warm
Sites

Paper Birch
Generalist

Figure 2.2. The images of species depicted in figure 2.1 are replaced here with descriptions of the species' habitat preferences.

stay cooler and therefore the soil stays moister. Thus, you can usually find plants that love dry soils on south-facing slopes and plants that love moist soils on north-facing slopes.

With this understanding of the differences between north and south slopes, take some time to stare at figure 2.2 until you understand the pattern in your body—words alone will fail. As expected, on the right side there is a suite of species that love warm, dry sites. On the opposite slope there is a suite of species that love moist sites. And look at the compass arrows. The red arrow points north, just as the dry slope on the right faces north away from the sun. The slope with the moist-loving—

Hold on.

But wait, did I say the north-facing slope is the hot, dry site?

And the south-facing slope is the moist site? I mean, that's completely opposite of what we'd expect, right?

Something's wrong here. I guess we'd better dig deeper.

When I confront a vexing problem in science or life, the thing that works best for me is to stop trying so hard to solve it. Instead of powering through with logic, I do better when I let my mind and body wander freely for a bit. This allows me to leap over to the framework where I really need to be. So that's what we're going to do now—wander. Trust me, it will all come back around to the puzzle at hand.

ARTIFACTS

While poking around Bark Hollow taking photos, I feel a wave of nostalgia as I look down to see a half-buried mechanical pencil, most likely dropped by one of my students years ago on a class trip. Picking up the pencil, I examine the dirt-caked sides and feel a fondness for this treasure. As I savor the memory of that class, my mind flips to a summer I spent scanning for ecology treasures with a metal detector.

Thirty years ago, out in California, Peggy Fiedler was crawling around the open hills of California counting individual leaves of the small, endangered Tiburon mariposa lily. This species grows only on the small Tiburon Peninsula, adapted to the harsh conditions of the local serpentine soil—starved of nutrients and loaded with heavy metals due to the underlying metamorphic rocks from which the soil is derived. In an effort to understand how this lily grew, Peggy used hundreds of toothpicks to diligently mark and number every plant in one-meter squares (about three feet by three feet). She returned for three years (the typical length of a graduate field study) to measure the growth of each individual plant and used these data to build a demographic model—a model that tracks how plants of a certain size or age grow and reproduce each year. Such models can theoretically be used to project a species' fate into the future. Around the same time, 3,000 miles away, Sue Gawler was doing the same thing for the endangered Furbish's lousewort, which only grows along a single dynamic river on the border of Maine and Canada.

Decades later Sydne and I set out to see how well the demographic models would mirror population dynamics over time. So we trekked to the hills of California and the banks of the St. John River and searched with metal detectors—at times in white full-body hazmat suits to protect against the encroaching poison oak—until we found, still stuck in the mud as Sue and Peggy had planted them, hundreds of tiny thirty-year-old toothpicks and tent stakes, along with a few other treasures, like a spent Civil War–era bullet deformed from impact. It's amazing how long things stay in place that have no cause to move—and how long our legacy can last in nature. What Sydne and I learned from the study wasn't so surprising. Models built on a mere three years' worth of data didn't say too much about a population in the long run, nor were they intended to—we really need decades-long studies to understand patterns in ecology.

SURVIVAL

Beyond mechanical pencils, what other sorts of legacies have people left in this forest? In the middle of the right half of the chapter-opening image, notice the big multiple-trunk tree. This particular tree is a red oak, but if we look around off the frame, we'll find many more such multiple-trunk trees of other species about the same size. What does this tell us? Normally, as an oak grows from an acorn into a tree, there is a single leading stem that grows up into a tall, straight trunk. But if you cut down that tree, of-

Figure 2.3. Multiple-trunk oaks suggest a historic logging event.

ten little sprouts will regrow from the rim of the stump—a growth form known as coppice. Some of these stems may then grow up into full trunks again, this time with multiple big stems coming from a single base (fig. 2.3). A forest of multiple-trunk hardwood trees, particularly when multiple species are involved, often indicates a legacy of past logging. But it's not always logging that kills the aboveground half of a tree.

Consider the American chestnut on the right side of the chapter-opening image. It's a cluster of leaves at the base of a small dead shoot. Across the East, chestnuts primarily exist as stump sprouts. Massive chestnut trees used to dominate our forests—it's commonly suggested that prior to the 1900s, one in every four trees east of the Mississippi was a chestnut, and in Kentucky and Tennessee, it's said that one in every two trees was a chestnut. The nuts were a primary food source for bears, deer, fishers, turkeys, squirrels, mice, and humans. But then in the early 1900s, a fungus brought over on ornamental Japanese chestnuts jumped to American chestnuts and rapidly decimated the trees. By 1940 most American chestnuts were struck. The blight only kills the aboveground part of the plant, and trees that are killed while still young are able to sprout new shoots from their roots.

I've read that in places, particularly in the Southern Appalachians, you can still find grand "ghost" chestnut forests, where huge stumps and logs of giant trees that have been dead a hundred years still dominate the landscape. I've long wanted to see one of these forests, and I'd really like to have a picture of one in this book. So this year I asked around and picked up some leads about places near the Great Smoky Mountains.

In April we headed to the Smokies. It was 4:00 AM when I snuck out of the Gatlinburg hotel room to check out my first lead, trying not to wake Sydne and the kids. Walking without a flashlight in the cold, moonlit stillness, I passed silhouettes of trees, some dead, on my way to the top of Ramsey Cascades. I noted the locations of the dead trees but figured I'd inspect them on the way down, once the sun was up. Halfway to the top of the falls, I entered the zone of old growth, full of ancient, twisted, towering trees that had never been cut. At sunrise I reached the cascades, 4.5 miles up the trail, then turned to head down.

Leaping from dead tree to dead tree, I peered into the wood end grain. As wood grows, minute patterns of lines and dots form in characteristic ways for each species. This log had prominent "rays" that radiated outward from the tree center—it must have been an oak. This stump still had bark on it—it must not have died very long ago. The concentric growth rings of this log didn't have the little dark pores that indicate water-transporting vessels—it must have been a softwood, probably hemlock. This one did seem a lot like a chestnut—a big, solid, hollow stump, rotting from the inside out like hardwoods do. But the pores in its wood were all uniformly small—it was probably a black cherry like the other big black cherries living around it. Nothing seemed to be chestnut.

Instead of finding the remnants of a century-old blight, mostly what I saw was another unfolding blight—the death of the hemlocks. They're being killed by a little fuzzy aphid-like insect called woolly adelgid, introduced from Japan. In places on my hike, the entire canopy was just standing dead snags, like a fleet of ship masts towering over a sea of flourishing rhododendron. The rhododendron was thrilled by the new bonanza of sunlight (fig. 2.4).

As I neared the bottom of the Ramsay Cascades trail, I found a few old logs and cut stumps that seemed to have the features of chestnut but were less than impressive. So I returned to the hotel in time for checkout.

Sydne, Juno, Alder, and I then headed back into the park to follow up on another chestnut lead. We parked the car and stepped out onto a ridge-top trail. Within a few feet we started seeing many dead snags, bleached and dried in the sun. These clearly seemed to be chestnuts; they reminded me of the dead chestnuts I've seen in New England. But they also weren't much bigger than the chestnuts I've seen in New England, and I felt a bit defeated. Juno was happy we found the chestnuts before flying back to the Northeast. But I still wanted to see something more dramatic.

Two weeks later the kids and I were back in Nashville for my high school reunion. The plan this time was for us to drive out to the border of Tennessee and North Carolina, where we would camp for the night. From there we could hike in to a very specific location in an old-growth forest where researchers had mapped out dead chestnuts in the 1990s. Their notes described it as a beautiful small ravine littered with dozens of chestnut logs, just uphill from a dense hemlock stand.

The forecast called for rain, but only in the form of showers. Whereas rain in the forecast means something persistent and often heavy, showers are usually fleeting and light. It was only like a 30 percent chance of showers at noon. If we got there at noon and then checked the forecast again, we should be able to wait for a break in the showers to hike in to the chestnuts.

Figure 2.4. Rhododendron flourishing under dying hemlocks in the Smokies.

At 6:00 AM I filled a couple jugs of water in the sink, stowed the gear in the car, and then loaded the sleeping children for the five-hour drive. We were headed to the area where the Olympic Park Bomber, Eric Rudolph, hid out from the police for years eating salamanders and acorns.

An hour into the drive, I realized that I forgot the sleeping bags. Thinking about my days in college when I used to sleep in a debris hut made entirely out of leaves and sticks, I told Juno we could just make a nest with our clothes and a couple of spare blankets inside the tent. Just the night before, in fact, we'd come across an old photo of me showing off one of my debris huts to my bemused grandparents. "Yes, Daddy, we will be birds!"

I then realized that I also forgot my rain jacket. Well, at least the children had rain jackets. Oh, and I forgot matches. That, we really needed to stop for.

In wilderness survival there's a list of priorities. Shelter, water, fire, food, in that order—at least as preached by the naturalist and cult hero Tom Brown. Shelter first, because you can die of hypothermia in a warm rain. Water second, because you'd die in four days without water. Fire third, because it keeps you warm and you can use it to sterilize water and cook food. Food last, because you can survive two weeks without food. If you're dropped naked in the woods, your most important tool for survival, as Tom Brown preaches, is a knife, which you'll need to carve your fire-by-friction set. It doesn't need to be a knife of modern steel; you just need a rock that breaks with sharp edges to fashion a stone tool.

We stopped at a Love's truck stop, got three lighters for 99¢, and drove on to the mountains. About an hour from the site, we found ourselves on an exceedingly winding road. Halfway through, a photographer under a blue event tent took a picture of us. I stopped to ask him what was going on. The road, apparently, is infamous. We can now go online to buy a photograph of ourselves driving there. This, the Tail of the Dragon, is one of the curviest roads in the world, and folks come from all over to ride their motorcycles on it. That explained the strange biker clubs with dragon images and sculptures that we passed. And the signs exclaiming "318 curves in 11 miles!" And Juno's favorite, the "Tree of Shame," with parts of wrecked motorcycles hung all over it.

As we pulled into the trailhead parking lot, the dashboard thermometer said 54°F outside, and it was definitely raining. I went to check the weather forecast and realized I had no cell reception. I'd had a smartphone for less than six months, after defiantly holding on to my 2005 flip phone for twelve years, and I was already dumbly dependent. When would there be a gap in the showers? How should we decide when to start hiking? I guessed we should just go for the hike and hope the showers would pass soon. Freed from the cell phone, we could now live in the present.

I packed our bag, got the kids' rain gear on them, and donned my dad's old jacket not meant for rain. I found a couple umbrellas stashed under the seat—I'd never before hiked in the woods with an umbrella, but this seemed essential that day. With Alder loaded onto the carrier on my back, we set out on the trail. Immediately, we were rewarded with a big, bright orange spring salamander, who seemed to be eating a slug in the middle of the trail. A few minutes later we found another bright orange salamander, this one a black-chinned red salamander. It seemed odd to stumble across such large salamanders in the middle of the day, but I guess it was a sign that this forest is really healthy and the streams are really clean. It was also a sign that it was really raining.

With Alder chanting "more 'manders!" these finds provided motivation for our rainy hike up the steep trail. A mile and a half up the trail, we got to the point where we could either break off to the chestnut site or wander another three-quarters of a mile loop trail among giant 450-year-old tulip poplars. I really wanted the kids to see the old trees, and maybe if we wandered, it would give time for the showers to pass. The trees were huge, the forest understory was open and park-like, the ground was littered with wildflowers, and the wet leaves glowed bright green in the foggy rain.

We touched the bark of the giants, the kids posed for photographs, then we paused for a snack huddled under the two umbrellas I held. We got wetter and wetter. It didn't feel like scattered showers. It felt more like heavy rain. We had better just get to the chestnuts and get back to the campsite by our car.

I glanced at my map, and we cut off the trail. The chestnuts shouldn't have been far from that point. But immediately, we encountered a tangle of thick rhododendrons. Rhododendrons are an awesome plant, fond of the acidic soils around here. But they can make an impenetrable wall.

I folded the umbrellas, directed Juno through a low hole underneath some branches, then climbed up to the top of a huge mossy log. With Alder on my back, I balanced my way across the first impasse to meet Juno. Fighting through shrubs, saplings, and last year's thorny raspberry stalks, we inched forward across a very steep slope. After ten minutes I looked at the site notes and realized we had started in the wrong direction, so we turned up the hill to correct. To prevent from slipping downhill, I pointed out a sapling a few feet upslope and told Juno to hang on to that while I pushed him up. We inched toward the chestnuts.

At last I saw up ahead where we needed to go. But it was too late. There was one big fallen beech just up from us. All we needed to do was to scramble over that log, then we'd be at the spot in the old field notes littered with chestnuts. But the thing is, I no longer cared about chestnuts. At some level I knew that this whole journey had a singular mission, which we were on the cusp of accomplishing. But I'd been hiking through the rain for hours with a one-year-old and a four-year-old, we were off-trail, two miles from the empty trailhead parking lot, with no cell reception, no other person within miles, cold and wet to the bone, exhausted from scrambling on slopes, and my shins were bruised from banging into branches. My body was moving toward survival mode. Shelter. Water. Fire. Food. Old rotten logs just weren't on the list.

Where we stood, Juno spotted some big logs covered in a thick carpet of moss. Noting the young saplings taking hold in the moss, he excitedly exclaimed "nurse logs!" and implored me to take a picture. Could these logs be chestnuts? Maybe, I guess. Whatever. I saw nearby some American ginseng, suggesting that it was a calcium-rich forest, relatively unplundered by people. Without moving my body I despondently aimed my camera at the logs and snapped some pictures, but my lens was foggy.

Even if we found chestnuts, I doubted my wet lens could get a clear image. I could picture in my head what the chestnut logs would look like; did we really need to go see them? I felt like we'd gotten close enough to at least get a sense of the forest. And one thing's for sure: this was no longer the same beautiful, forested ravine that the researchers had visited in the 1990s.

Our off-trail trek was difficult in large part because the understory was much thicker than it was twenty years ago. Why? Because all the hemlocks are dead from adelgid. The beech, too, was likely felled by the recently introduced beech bark disease. Whereas the nearby tulip poplars haven't yet been struck by an invasive pathogen (knock on wood), this little ravine had seen its canopy of chestnuts, hemlocks, and beech all destroyed in the last hundred years. It was a stand in transition full of young growth.

The kids were still in great spirits but also ready to go. Going over that beech log just felt like one step more than reasonable. This, I guess, was my limit. I asked Juno if he'd like to go back to the trail or keep going. When he said he was ready to go back, I decided not to push him any further. More than anything, I wanted the kids to enjoy this adventure. Someday we'd have to return on a warm, sunny day. "No!" Juno protested. He only wanted to come back in the rain—to see the salamanders again.

So we headed back down to our tent in a valley, with the kids pretending to be motorcycles on the Tail of the Dragon. Behind our tent stood an open hardwood forest of tall tulip poplars, maples, and blooming dogwoods. In front of our tent, the opposite slope was blanketed densely with hemlocks and rhododendron. I wondered, was this pattern driven by the same forces driving the parallel vegetation pattern at Bark Hollow?

Back in the Northeast, I relayed my unsuccessful attempts at finding chestnuts to my colleagues. They weren't surprised. It seems that part of the problem is that the historic role of chestnuts in our forests may have been overstated. Ed Faison and David Foster did a recent analysis of old "witness trees." These are big trees that were historically left uncut to mark property boundaries. These trees can give us some sense for the original forest, prior to large-scale deforestation. Rather than 25 percent to 50 percent of the original forest, the witness tree data suggest that chestnuts were only 5 percent to 10 percent of the overall forest, if that. Although in local forests chestnuts were much more dominant.

Then an email from ecologist Tom Wessels sent us to the shoulder of a mountain in the very corner of Massachusetts. Here, as he described, was a thirty-acre stand with chestnuts still surviving up to a foot in diameter, getting big enough to cross-fertilize and produce viable nuts before succumbing to the blight. Here a thick wall of mountain laurel covered the slope—filling much the same niche as the rhododendron in the Smokies. Above the mountain laurel, it wasn't the grand forests of towering chestnuts of the past. But when Alder, Juno, Sydne, and I made our way up the mountain, we found chestnuts everywhere, including many seedlings along with many larger individuals much older than any I'd ever seen (fig. 2.5).

Somehow the trees in this stand were managing to just barely hang on to the thread of evolution. As long as they can keep re-

Figure 2.5. American chestnut: (a) a snag in the Smokies, and (b) a living individual, old and twisted from battling the blight and barely surviving to reproduce, in a stand with many similar trees in Massachusetts.

producing sexually, and if there is enough genetic variability left in the stand, generation after generation of natural selection will mix up their genes until nature arrives at a solution to the blight. It's places like this, Ed Faison says, that are his hope for chestnut's future. With chestnut seedlings seemingly sprouting all over, this corner of a mountain is nature's little laboratory. Scientists elsewhere are scrambling to produce hybrid American-Asian chestnuts, GMO chestnuts infused with a gene from wheat and virally weakened strains of the blight. But, ultimately, nature on its own may be more effective at chestnut restoration across the landscape. It will just take a long time—hundreds of years.

Back in Bark Hollow, could human activities such as logging have caused the left-right pattern? Or, similarly, could the pattern have been caused by a disturbance such as fire or wind that knocked down half of our site? Put another way, is one forest a lot older than the other? If so, we might see differences in the sizes of trees or the relative proportion of big versus little or understory versus canopy trees. A younger, or even a much older, forest might also be dominated by fewer tree species, rather than having a mix of many species.

Examining the chapter-opening image, the size of the trees and the relative proportion of big and little trees seems about the same on both sides. And neither side seems dominated by a particular species—rather, the two sides show a good mix of deciduous species, even if there aren't the same species on both sides.

But I do see evidence of humans currently using Bark Hollow. Though we got here off-trail, we seem to have arrived back at a trail of sorts—right down the center of the image. It's wide enough that people, likely even people on motorized vehicles, occasionally saunter through. And people probably aren't the only ones using this trail. Animals, like people, are always looking to conserve energy—you might say we're all naturally lazy. A cleared trail through the underbrush or snow is easier for animals just as it is easier for us. One question that's always good to ask about a trail is, why is it located here?

From where we're standing, if we look either to the east or the west, it's downhill in both directions—water would be draining away from us either way. If we look to the north or the south, it's uphill in both directions. We are standing at a tiny "saddle." Two natural travel paths intersect at saddles—the high point for valley-goers and the low point for ridge-walkers. Animals walking along wetlands in the valleys to the east or west get funneled into this saddle point—it's the easiest way to get from one drainage to the next. If you're interested in walking along the ridgetops, you similarly get funneled down into saddles. Wherever trails intersect is a place that animals—particularly predators—hang out to smell who's been by and scent-mark to leave their message. Thus, a saddle seems like the kind of place I would look for signs or put a wildlife camera if I wanted to see who's around.

Visualizing how animals will use the landscape is a fun game. If you're a beginning player, try river otters. Otters spend most of their time in ponds, lakes, and rivers. But they also move over land back and forth between water bodies, usually selecting the quickest route. If you look at a map, you can often pinpoint exactly where the otters should cross from one pond to the next. Then go out and look, and you may find a well-worn otter highway, sometimes marked at the ends by scat and otter scent piles and sometimes with a groove worn into the earth itself from the heavy use, even if the trail is a half-mile long.

So the trail is here in our site because it was funneled into this saddle. But, as Juno might ask again, *why*? Why is there a saddle here? Answering this requires delving into some geology.

PLATE TECTONICS

North America keeps bumping into things. Over the past billion years, it's happened several times. Each time, the leading edges of North America and the offending landmass buckle and crunch like cars in a demolition derby. And then, like a feuding couple, the continents eventually separate again—each carrying fresh scars. Rhode Island, for instance, used to be part of the drifting microcontinent called Avalonia before it glommed on to North America 450 million years ago—just as plants and insects were beginning to leave trilobite-infested waters for land.

Every time North America is squeezed from the side, like a water balloon in your closing fist, the top of the land rises up higher into the air. In this way, collision after collision, the Appalachian Mountains were formed. When amphibians were the dominant land predators and ginkgoes were just getting started, about 250 million years ago, the Appalachians were the size of our modern Alps.

Deep under these towering mountains, not far from our field site, a pocket of molten rock was slowly cooling. In this cooling magma, solid mineral crystals began to form—most notably quartz and feldspar, along with a few friends like mica and amphibole. The slower the cooling process, the more time the mineral crystals would have to form, and thus the bigger the crystals would grow. As dinosaurs began ruling the land above, the crystalized minerals cooled into a mass of solid rock. This rock, made of coarse grains and dominated by quartz and feldspar, would later be called granite by the naked apes.

But just as our little pocket of granite rock under the mountain was forming, with pterodactyls gliding overhead, the rocks on top of the mountain slowly weathered down. Big rocks broke into little rocks from the mechanical forces of wind, rain, ice, fluctuating temperatures, and stomping dinosaurs. Not only did rocks break, but the mineral composition began to change too. Although minerals like feldspar are stable deep underground, when they get to the earth's surface, feldspar doesn't last long. Such minerals chemically disintegrate and transform into more stable minerals like quartz and clay, along with dissolved ions carried away by water. Given enough time at the surface, most rocks will eventually turn into fine-grained quartz and clay. But not all rocks are given the time they need—sometimes they're reburied and rejoined with other rocks. Loose gravel, incompletely weathered and piled deeper and deeper at the foot of a mountain, might get transformed into sedimentary conglomerate given the right conditions.

While plesiosaurs chased Triassic fish 200 million years ago, the new Atlantic Ocean was about to form. Africa and North America were preparing to split. As the two continents set off on their separate ways, the land along the border stretched, thinned, and began to break up. In our little corner of the continent, a spreading fault line became a sinking rift valley as the mountainous walls of the valley moved away from each other. Repeated earthquakes tilted the valley floor as it slipped downward.

Molten lava erupted through the thin skin of the earth in the center of our little spreading valley, pooling in a huge, fiery lava lake. In the lava, characteristic minerals like olivine and pyroxene were forming. But it cooled and solidified relatively quickly, without enough time to form large mineral crystals—or even crystals big enough to see—into a rock known as basalt. Whether you find it in Iceland, Yellowstone, or on the ocean floor, basalt is a rock to know, as it reveals a volcanic past. While not as glass-like as obsidian, when broken, basalt still reveals relatively sharp edges due to the lack of large mineral crystals inside. Because of this, in a pinch, basalt can make a good primitive blade. And unlike many other rocks, when chemically weathered, basalt will readily release cations like calcium, sodium, and magnesium.

Volcanoes and earthquakes all promised to turn this little valley into a grand ocean separating two great continents. But promises aren't always fulfilled.

Up and down the east coast, from South Carolina to Nova Scotia, many such rift valleys were all dreaming of becoming the new Great Ocean. The winning contestant was a rift valley one hundred miles east of our little inland valley. While our valley tried to send Boston and Rhode Island off with Africa, it was not to be. The Atlantic Ocean started spreading, as it continues today, with the new ocean floor spewing out from the underwater volcanic mountain range of the mid-ocean ridge. As our mountains still crumbled into the valley, bursts of volcanic activity slowly subsided and stopped here.

According to some reconstructions, when *Triceratops* and *Tyrannosaurus* arrived, sixty-eight million years ago, they saw nothing of the great mountains and volcanoes that once marked our valley—not even the valley itself. Whereas eighty million years prior, *Stegosaurus* and *Archaeopteryx* had frolicked in the mountainous scenery, time had ground down the mountains and dumped them as fill into the valleys. Now there may have been just a gentle plain sloping down toward the new ocean.

But then as mammals began to dominate and only little birds remained of the great dinosaur lineage, the edge of our continent began to lift upward. Driven by tectonic forces below, the landscape was raised hundreds of feet into the air. Water in streams now fell further on its way down to the ocean, cutting with more force through the underlying bedrock. As dog-sized primitive horses galloped along the riverbanks, rushing waters pounded down through rock history. The water revealed layers of basalt from volcanic eruptions stacked between intervening sedimentary layers—the crumbs of the old mountains. And because our valley had been tilted millions of years ago during the rifting phase, these layers were angled sideways, with their edges now protruding from the ground. Basalt, it turned out, was harder for the rivers to cut through than the sedimentary layers. As sedimentary layers were washed away, basalt mountain ranges—the edges of the old lava lake that had hardened and tilted—rose up higher and higher from the valley floor.

It was a slice of that lava lake that I carried Juno up this early August morning to get here at Bark Hollow.

HEPATICA

Wrapped in my childhood blanket, Juno falls back asleep in the middle of our site. I take a breath and wander over for one last look at the hepatica. For such a small plant, hepatica seems to have quite a power over me. Why?

Is it that leaf shape? Bold, defiant, unapologetic. It didn't choose the standard three-leaflet form—like poison ivy, jack-in-the-pulpit, tick trefoil, bladdernut, or any number of plants where each leaf has three distinct blades on the stem. Nor do its three conjoined lobes accept the standard forward-thrusting, turkey-foot-like form that sassafras, great ragweed, and mulberry chose. It feels sinister, the way hepatica's symmetry slips through the forms my mind wants it to fit.

Is it the name? From the Latin *hepaticus* because it resembles a liver. Does some part of me think I see my own entrails in the leaf litter? My fear of death seems at hand with this plant. With cautious approach, I don't stare too long, just in case, as my primal reasoning goes, I might acquire some disease through association.

Is it the medicinal properties? I've never tried hepatica. I hear it's neither good tasting nor good for me.

Despite the dark associations in my mind, hepatica actually carries a strong message of hope. In late August these thick, fleshy leaves are a reminder of the early spring wildflower show. While most of the wimpy leaves of the spring ephemerals have long since withered, hepatica leaves will persist all winter. Hepatica's dainty lavender flower is one of the earliest blooms of color in spring. Closing my eyes, the rest of spring floods into view with carpets of toothworts, spring beauties, trout lilies, and squirrel corn amid the joyous birdsong and thawing air of an awakening forest.

More than just a time capsule to spring, hepatica makes a statement about the land itself. Where you find hepatica, you also often find a lot of other special plants—like wild ginger, ginseng, Dutchman's breeches, and rue anemone. Plants that in many regions are often rare or under threat. Plants that remind me of the rich central Tennessee woods where I grew up, on top of limestone.

These rich woods teem with life, and that's what I sense when I see hepatica.

Little hepatica. A small exclamation mark tucked into the leaf litter, the lone remains of a rich spring statement. To reconstruct its meaning you must have seen such punctuation in context. If not, your eyes scan past, your feet glide over, and you don't hear hepatica shouting.

CAPTURING THE ROCKS

When we return home in the afternoon, I examine the pictures I've taken. I flip in frustration through the mediocre images on my computer monitor, and it is clear that I failed to capture all the pieces. Most significantly, I need better shots of the rocks.

A whole year passes before I set out again to finish photographing Bark Hollow. I awake in the morning, eager to leave the house. The first stop will be to drop Juno off at preschool, and this time it will be Alder who will come with me up the mountain. I admit that some of my eagerness to get out the door is because I have filled my car rides this week by listening to an audiobook, *The Magpie Murders*. I'm hooked. Can I assemble the clues before the narrator reveals the solution? What better guide to writing my own nature detective book than a murder mystery?

As we leave the house, I pause at the large pile of heavy rocks outside the front door. In fact, these rocks have their own murderous record—they once nearly killed me and a colleague.

I bend down to hold a few of the rocks now, remembering the journeys they have taken. There is the flat, reddish chunk of mudstone, two feet across. If you search this layer enough in the field, you'll find abundant traces of insect tracks and occasional dinosaur footprints from 200 million years ago. There is the seventeen-pound chunk of granite. The zebra-striped gneiss. The colorful, coarse conglomerate.

And then there is the finer-grained sedimentary rock. This rock comes from a rock unit that is primarily an arkose sandstone—made of sand-sized particles whose high amount of feldspar shows they haven't been fully weathered. However, toward the eastern edge of the rock unit, the particle sizes grade into something larger. This chunk of rock sitting before me has pebble-sized particles and so would be classified as a conglomerate within the arkose layer. This layer was made when the ancient mountain crumbled into the valley.

It is the arkose layer that I'm heading to photograph this morning. Instead of climbing the mountain with child in tow, why not just take a picture of this rock here in my yard? My lower back would certainly prefer that. After all, hadn't I collected the rock from the very same spot at Bark Hollow? Alas, I am a purist. What if Atticus Pünd, the sharp-eyed detective from the *The Magpie Murders*, ever reads my book? He will surely note the shadow of the yew bush that doesn't belong.

And there in my yard is the smooth black basalt—from the ancient lava that flowed into the rift valley. For my book I wanted to get a picture of this basalt turned into an arrowhead. So I recently grabbed a big chunk from this rock pile and went with Juno to see Neill Bovaird—an old friend from the Woodsy Club.

Standing in Neill's yard, I asked him if he could make an arrowhead or spearhead from this basalt. He said he could, but that's not really what basalt was traditionally used for. Arrowheads, Neill informed me, would have been made with local quartzite perhaps, or, more likely, flint traded from upstate New York. But the basalt layer at Bark Hollow? That's where everyone would go when they wanted to make an axe or an adze.

In a few deft strikes, Neill broke the rock into five fragments, with piercing thuds, and they fell to the grass. He was left holding one small piece, which he tapped on. "Hear that?" This piece of the rock made a higher-pitched sound, indicating that it didn't have any cracks in it. Those other pieces all had cracks. This one he could make something usable out of.

We left the shards outside and walked into Neill's house. His workroom upstairs was full of natural crafts—arrows, bows, bowls, baskets, skins, skulls, fire-making sets. Juno stood in awe of it all. Neill showed off a couple recent projects he was proud of: beautifully ornate, two-toned woven dogbane baskets and soft and tough fish skin he'd leathered. And then there were all the stone tools. Neill picked out a few basalt pieces in different processing stages. One was a Native American artifact a friend of his found. Another was a finished axe-head. For the best photograph of the axe-head he would make from the rock we'd brought him, he suggested that he mount it on a handle. Come back in a week or so, he said.

A few weeks later, I found Neill at the nine-year celebration and fundraiser for the camp he founded, Wolf Tree Programs. In a forest of slender trees stretching up to the canopy, a "wolf tree" is one whose massive trunk and horizontal lower branches tell that it grew here long before all the others, alone in a field. It was a cold rainy day at the celebration, but the community turned out, and it was a lively event.

On a table in the center of the celebration, I found an impressive display of crafts from Neill's workroom. Between a box turtle shell, a plaster track cast, a pair of deer antlers, and a finished arrow was a freshly carved piece of wood. The elbow at one end told that the wood came from a branch joint on a tree. At this bent end, bound with sinew, was the axe-head Neill shaped from the basalt of Bark Hollow (fig. 2.6).

Still staring at the pile of rocks in my yard as Alder and I prepare to revisit Bark Hollow, I think of why I gathered all these rocks—basalt, mudstone, conglomerate, granite, schist, gneiss. On the first day of my *Field Naturalist* course, students arrived to

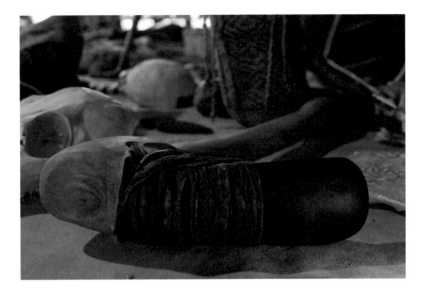

Figure 2.6. Adze made from the basalt of Bark Hollow.

find these very rocks scattered across the tabletop. Hand-selected from each of the sites we would visit throughout the semester, the rocks told the students of the grand adventure ahead of us. The students' job that first day was to arrange the rocks into groups. I gave them no information about the rocks' origins or names, and most students had no background in geology. Yet the rocks slowly made their various ways across the table toward each other, finding appropriate geologic groups. The sedimentary rocks with the aggregate bits of particles. The igneous rocks with their randomized crystalline structure. The metamorphic rocks with the crystals formed into wavy lines. At the end of the exercise that day, without revealing any answers, I merely packed up the rocks and put them up on a shelf in my office to sit on display until needed again. The students then came to discover the same rock types later in the field, like finally meeting the parents of an old friend.

It was putting the rocks on the shelf that caused the trouble.

High above my desk, the shelf, it turned out, wasn't fully fastened to the wall. Many weeks passed before I came into my office for a routine meeting one morning. Not long before I arrived, people in the hall had heard an enormous crash erupt from my office. The shelf, along with the great weight of the valley's rocks, had leapt across the room. Wood had splintered from the table, the floor was dented all over, and many of the rocks had split. I shoved the rocks to the sides, and a few minutes later had my meeting. I'll never forget the uncomfortable look of my colleague, sitting amid a sea of rubble and the smell of fresh rock dust, knowing that we had both almost been smashed to bits. Afterward I added some extra supports and put the rocks back on the shelf, though knowing visitors always remained nervous.

NUTRIENTS

Finally back at Bark Hollow, I watch from up on the slope as Alder, several yards downslope, practices standing by leaning against a large, bumpy rock. Alder was named after the small wetland tree, a tree known for its important role in natural communities. Through a symbiotic relationship with nitrogen-fixing bacteria on its roots, alders fertilize the soil. These enriched soils set the stage for subsequent species to flourish. Here in this field site my son is the only alder growing. But there is something else giving nutrients to the soil.

Alder's patience with me wandering around the site fades, and I bound back down to scoop him up before too many tears are shed. I note the rock he's been leaning against. It's the same rock type on the bottom of our chapter-opening image to the right of the compass. It's irregular and seems to contain a mishmash of smaller rocks of various sizes and colors. These interior rocks are themselves variously weathered—with rounded edges, not sharp crystalline edges. Thus, we have a sedimentary rock—made from loose pieces that were compressed into a solid mass.

This sedimentary rock is quite unlike the spectacular red Navajo Sandstone pictured on Utah license plates. The smooth Navajo Sandstone is made mostly of quartz sand that was blown into a well-sorted pile of uniformly small, round grains. Weathering and sorting happen when wind and water are given sufficient time and distance to work.

No, our sedimentary pile seems to be very poorly weathered and poorly sorted. Instead of purely quartz, we still find a lot of feldspar along with other minerals present. Instead of uniform particle size, we see many different sizes. This pile was evidently formed not too far from the large mountain it came from, buried soon after it was deposited. The feldspar tells us of the granite in the mountain that once was. This is the conglomerate from the arkose layer.

I want to test the soil here, and Alder helps me dig into the ground. I forgot to bring a pH kit today, so we need to take some samples home to test later. We scrape our knuckles in the gravelly soil, looking for places with few rocks to get a good handful of fine dirt to test (fig. 2.7). What should we carry it back in? Searching my backpack for something to hold the sample, I find a weeks-old half-eaten egg-and-cheese croissant that Juno had asked me to save for later. The croissant is wrapped in foil. I tear off some foil and we pile in some soil.

I close my eyes and think about the giant mountains that once stood here. Today our closest real mountains are New Hampshire's White Mountains. I think of Olivia Bartlett, who is earning her PhD by scrambling up and down those mountains studying the soils.

I recently showed Olivia a picture of our little Bark Hollow, and she keyed right in on the basswood. On her mountains basswood is a strong indicator of the soil chemistry—where she finds

Figure 2.7. The rocky soil in Bark Hollow made of weathered basalt.

basswood, the soil has a lot of calcium and magnesium. Eager to understand soil science, I began interrogating Olivia about the term "rich."

"Rich," when applied to soils, refers to the availability of essential plant nutrients, like nitrogen, phosphorus, potassium, calcium, and magnesium. Calcium holds together plants' cell walls.

Magnesium serves as the central atom in chlorophyll. Nitrogen and phosphorus are central components of DNA. Potassium controls leaf stomatal openings. All of these nutrients serve many other functions, and many other nutrients are needed for plant life.

My question to Olivia was why there seems to be an association between richness and moisture. It's not that all rich places are moist, nor are all moist places rich. But often plants that do well in one do well in the other. All else being equal, my sense is that if you add moisture to a dry habitat, it will support more species that love rich soils.

What I wanted Olivia to say was something simple like, "Yup, in moist soils, the extra water helps break down the parent rock material faster through chemical weathering, thereby pulling more nutrients out of the rocks and into accessible forms in the soils." But that's not what she said. Soils are complex, and it's not simple to predict soil characteristics. Soils are formed by the breakdown of rocks and sediments below the surface and the decomposition of vegetation above. These processes are in turn influenced by geologic forces, climate, topography, the work of burrowing animals, the amount of time that has passed since the last disturbance, and the movement of wind, water, and ice.

In Olivia's work, for instance, she sees that soils develop very differently depending on whether the groundwater at a spot flows horizontally or vertically. And consider plowing. Natural soils typically have distinct layers of colored bands as you dig down (called "soil horizons") with different properties. At the top there's typically a dark A horizon full of rich humus—decomposed organic matter. But in fields that people plow, the topmost layer gets all mixed up with the mineral layers below, and this can have a profound effect on soil nutrients, even centuries later.

To further complicate things, as soils become too acidic or too basic, the nutrients become less available to plants. Too many hydrogen or hydroxyl ions floating around cause the nutrients to bind up into compounds that the plants can't use.

Back home with a cheap pH kit from the garden store, I discover that the soil on the left slope in our chapter-opening image measures around 5.5, whereas the soil on the right slope measures around 4.5. In our region of New England, soils tend to be acidic. A pH of 4.5 is pretty typical.

When we have the occasional ancient lava flow driving soils into a more neutral range, it's special. That means we're going to find a lot of plant species growing here that won't be common across the rest of the landscape. If we travel to a limestone-rich region like middle Tennessee, we might see the reverse pattern. There we might expect to find the rarest plants on the landscape in occasional acidic soil islands amid a sea of richness.

Reflecting back on my question about the connection between richness and moisture, Olivia asked whether we might just be finding richer soils lower down on slopes because groundwater is transporting the nutrients through the soil from high points to lower points. Or is it just that the rich soils I'm thinking of are in the floodplains of waterways that deposit extra nutrients in the soils during floods? I guess the answer is, it's complicated and nobody yet has a full understanding of how soils work.

But we do know a little. We do know that basalt—the dark, angular, fine-grained rock on the left side of Bark Hollow—releases a lot of magnesium and calcium and raises the pH of soil as it weathers. That's not the case for the rocks within that arkose layer, which we see on the right side of the image.

And so we have the source of our pattern. The path in front of us delineates the edge where two bedrock layers meet. Hepatica and the other plants on the left are growing in rich soils formed from weathering basalt. The plants on the right are growing in the relatively poor, arkose-derived soils. Like Tiburon mariposa lily growing in the serpentine outcrops of Marin and like rhododendron's affinity for acidic slopes in the Smokies, the key is in the soil chemistry. It's not the moisture preferences in figure 2.2 that are important, it's the plants' preferences for soil richness. Bark Hollow is the crack between two layers of geologic past, slowly breaking down. Our plants are telling us the story of when Africa left our continent 200 million years ago.

FINDING THE SPOT

It was Karen Searcy who sent me to the south side of this mountain. I had looked at United States Geological Survey (USGS) bedrock geology maps and found arkose-basalt junctions on the north side of the mountain. But, Karen told me, the plants only reflect the pattern well on the south side. Karen should know; she spent years studying this pattern. My friend Meggie Winchell, one of Karen's students, describes sitting for hours staring at aerial photographs on a computer and circling patches of mountain laurel. Then Karen and her team fanned out to points all over the landscape, probing the soils and identifying the plants around Meggie's circles. The pattern we saw at Bark Hollow repeated over and over across the landscape.

But it's rare to see such dramatic bedrock-driven patterns in the Northeast. Over the broad scale, yes, you tend to see more rich-loving plants in places like the Berkshires that have a lot of marble. But these patterns are often masked, especially at the small scale. If you travel down south, you'll see plenty of bedrock directly influencing the soil. But in the North the soils often don't reflect the bedrock below. For that, glaciers are to blame.

A mere 20,000 years ago, Bark Hollow was pinned under a mile-thick sheet of ice. As the climate warmed and the ice melted away, it dropped all the sand, gravel, rocks, and boulders it was carrying. Today most of the Northeast is covered by a five-foot-thick random jumble of various sizes and types of rocks. In addition to this "glacial till," the glaciers left behind other traces too. For instance, as the glacier retreated, the rift valley around Bark Hollow filled up with an enormous glacial lake, still evident from other surface deposits left behind.

Across the landscape the depth of glacial till is not uniform, and it's controlled by complex processes I don't understand: the patterns of glacial flow, ice melting, and subsequent erosion. Bark Hollow sits on the steep south side of a small mountain range, and this whole side of the range is largely free of glacial debris. Perhaps it's the steep slope that caused the sediments to wash downslope as the ice melted. Whatever the reason, the bedrock is exposed here, thankfully.

MAJOR LESSONS FOR INTERPRETING A LANDSCAPE

- Consider the geologic history of your site. How is the bedrock shaping the soil chemistry, and how are the plants reflecting this?
- Learn to orient yourself on a map and think about the topographic setting of your site.
- How have introduced pathogens changed your landscape?

3. Water

Twenty miles from Bark Hollow, we drive northward along a straight country road through the woods. The land to the left is nearly flat, with a very slight tilt down toward a little facility in the forest. That's where state employees raise fish to stock regional streams and ponds. Looking to the right is a long and steep slope up—we are driving parallel to the base of this slope. We turn right, climbing upward. The road transitions from pavement to dirt, and suddenly we've reached the top. The land is level again, though the road is pocked with car-sized puddles that we weave in and around, splashing up and down. The kids in the back seat giggle at the bumps and splashes. We park the car and hop out onto the sandy road. After getting our gear together, we head off into the brush. We fight through thick branches and leaves for about a hundred feet before we arrive at our destination: Maggie's Forest (figs. 3.1 and 3.2).

As Juno gorges on blueberries and Alder naps in the carrier, I begin to snap photos. I'm immediately frustrated. This isn't how we left the woods four years ago. The pattern in the trees and plants that the *Field Naturalist* students set out to describe was blaringly obvious—though the cause of the pattern was not obvious. Now it's hard to find, and even harder to photograph. Though the tree pattern is subtle, there is some good news here. The changed forest

200 Yards ▲ N

Figure 3.2. Aerial photograph from 2005 with the black triangle indicating the location of this chapter's site. (MassGIS, Sanborn LLC)

now makes another feature of the landscape strikingly clear: the shape of the land itself.

Did you see these patterns in the chapter-opening image?

Like at Bark Hollow, the trees on the left are entirely different from the trees on the right: pines versus oaks. Unlike Bark Hollow, the land is flat. Really flat. Look at that horizon line. Why is it so flat? And what's driving the tree pattern? And why does the bottom half of the chapter image just look like a mixed-up jumble of vegetation?

Did I mention how flat it is here? If you were tasked with shaping this landscape, how could you make something so flat? I mean, if I set you out in the woods and gave you only the tools of nature, could you come up with something perfectly flat? I'll give you some milkweed to make primitive string, and here's some white cedar and mullein that you can rub together to make fire. Maybe you could smash up some quartzite for cutting tools, which, together with the fire, you could use to carve out a little wooden bowl. Drop some pine needles into the bowl along with some water and some burning hot rocks from the fire to boil your tea. If you're really ambitious, maybe you could dig out a whole log and float a little canoe to the far side of the lake where the tasty cattail stalks grow. You'll need to gather sticks and leaves to make a little debris hut for sleeping. Where in your day of hard work with these rugged implements can you find something flat?

PINE NEEDLE TEA

As we think this through, let's have a seat by the fire, munch some cattails, and sip our pine needle tea. I love pine needle tea. OK, I'll concede that it's got a bit of a turpentine flavor to it. But in a good way, I say. Besides, it packs a ton of vitamin C. You've never tried it before? Well, I've made it with all sorts of pine species while I'm out on various adventures. It's always a reliable go-to soothing drink. Oh, that's a good question, what species is this one? Let's see. Looking closely at the needles, it seems they are grouped in bundles of three. That is, there are three thick needles coming out of one base. That means it's not white pine—that species has bundles of five thinner needles, which give the whole tree a fluffy aura about it. It's the five united needles that gave white pine the nickname "Tree of Peace," representing the original five nations of Haudenosaunee that long ago came together under a peace treaty. And let's have a look at that pinecone—see the little pointy tips sticking up at the end of each scale?

Pitch pine. *Pinus rigida*. If you're looking for indicator species, here's a good one. Pitch pine grows on poor soils that are sandy or gravelly, and it is well adapted to disturbances such as fire. Unlike most other pines, if the aboveground portion is killed, the tree will resprout from its roots. It's so good at sprouting new shoots

Figure 3.3. A pitch pine exhibiting epicormic sprouting.

Figure 3.4. Winged seeds of pitch pine on a fire-scarred piece of bark.

after trauma that, unlike other trees, where the bark is damaged, new pine needles will readily grow from the side of the trunk—called epicormic sprouting. This makes for some occasional funny-looking trees with a fur-like covering of pine needles along the trunk (fig. 3.3). The thickness of the bark also helps shield the living cells inside the trunk from the heat of the fire outside. Some pitch pines—mainly those that grow in areas with a long fire history—produce cones that stay tightly closed on the trees for a long time, held shut by resins. These serotinous cones open when they are exposed to heat. A fire, clearing the ground of competing vegetation, warms the cones. The cone scales spread out like fingers on an opening fist, allowing many little winged pine seeds to float away, sprinkled by the wind across the earth freshly prepared for a new generation of pitch pines (fig. 3.4).

Pitch pines are a common sight throughout northeastern coastal areas like Cape Cod, Long Island, and much of New Jersey. In these so-called pine barrens, along with pitch pine you'll see a lot of the same species we find here at Maggie's Forest. You'll see dry-loving scarlet oak—that's the canopy species on the right of our chapter-opening image. You'll see dwarf chestnut oak—which lives in the understory here. You'll see bear oak—the thicket-forming shrubs we fought through on our walk in. All of those and more are easy to find in coastal regions. But the nearest beach is one hundred miles away from here, and you'd be hard-pressed to find many of these species in the intervening landscape. What does this little inland site have in common with the coast? How did this pitch pine forest come to grow here?

Well, perhaps "forest" isn't the best word for what we're looking at. I suppose this looks more like a savanna. The trees are spaced far apart, and you can see lots of sky between the crowns. Forest canopies are typically closed, shading most of the ground. Where tree growth is restricted by lack of water, frequent fires, or heavy browsing, the tree canopies don't meet, and you get a savanna. Lots of sunlight reaches the ground in savannas, and beneath the trees you get a dense layer of grasses and other herbaceous plants (fig. 3.5). Think Serengeti, where the giraffes browse on tall acacia trees and zebras graze the grass.

But unlike the Serengeti, the understory here isn't grassy. No, it's shrubby. It's dominated by woody vegetation. In fact, the plants in the understory are primarily tree species—in the foreground of the chapter-opening image there are lots of oaks. This is a system in flux. With so much light available coming down from above, these oaks wouldn't be content just hanging out at toddler height. They are racing as fast as they can up to the canopy. Their small size can only be explained by the fact that they just started climbing in the last few years. Poke around under the shrub layer and you will find many stumps of recently felled trees. There used to be

Figure 3.5. Globally rare oak savanna ecosystem at Indiana Dunes National Lakeshore.

a dense tree canopy above, holding the understory oaks back. But then: *chop*, and the oaks were released from the shadows and began climbing up to the light. And some of these growing stems are shoots resprouting from the remaining bases of the old cut trees.

I remember those trees; we were standing in a closed-canopy forest when I taught my class here four years ago. We brought with us an increment borer—the tool foresters use to extract pencil-thick cores from trees for examining the rings—and the students proceeded to core the biggest canopy trees on both the oak and pine sides of the forest. Both were about a hundred years old. But the quick work of chainsaws brought an end to that hundred-year-old forest. Before I rile myself up with anger at how they destroyed my trees and my forest, I should remind myself that this land is owned and managed by the state wildlife agency. This logging operation was designed to improve habitat for rare species and as a component in the state's broader conservation objectives. If you're fresh out of the black-and-white world of environmental activism defined by the Lorax, the fight over clear-cutting of Pacific Northwest old-growth forests, and emotionally charged online petitions, this idea might take a minute to get used to: cutting down trees to help the environment.

FOREST SUCCESSION

What happens when you cut down trees? To start, we see logging as a "disturbance." Cutting down trees disturbs the forest. The trees have been standing for decades or centuries, creating a cool, shady, moist, tranquil shelter for the plants and animals that live beneath them. When machines come in, hack up the trees, and haul them away, everything changes. Sunlight beats down. Winds cut through. Some plants and animals are trampled during the operation; others can't survive the new hotter and drier environment.

But logging is just one type of disturbance. Fires. Floods. Wind bursts. Tornados. Hurricanes. Ice storms. Beavers. Insects. Even just trees tipping over from old age. Disturbance is natural. Sure, intact forests are important habitat that many species depend on. However, forests are but one type of habitat. Post-disturbance landscapes offer other types of habitat that different species depend on.

To really understand forest disturbance, we need to first get our heads around the language of successional stages. Imagine a forest dominated by mature oaks and pines. At the edge of our imagined forest, there is a freshly plowed farm field covered with dirt ready for planting. But instead of planting her crops this year, the farmer at last decides to move out West, abandoning her field. You see, she doesn't have enough firewood to keep her family warm for even one more winter (this is the story of many New England farmers in the mid-1800s). She's cut the last of the trees in the woodlot she owns. Perhaps decades from now there will be a new forest on her land sufficient to support a homestead. She wonders, will it be an oak forest or a pine forest?

After she heads West, first some "pioneer" plants move into her field. Small, fast-growing species whose seeds spread easily by wind. Grasses, dandelions, goldenrods, milkweeds. For several years these short-lived plants dominate the field. Cottontail rabbits hop from tasty morsel to tasty morsel. Meadow voles tunnel through the grasses. Toward dusk a red fox, seeing vole on the menu, trots out to the middle of the field, cocks her head, and listens. For a few minutes all is still. Then a soft flutter of tiny feet come racing through the grass beneath her. The fox pounces and, after a little tussle digging through the grass and dirt, is rewarded. She gulps the meal, leaves a little spot of blood, places a proud mark of her own urine, then trots off. As night falls, coyotes come to join the hunt for small mammals, and deer cautiously graze. Things might have gone on like this, day after day, forever. But the trees in the neighboring forest start sending in their seed missionaries, promising to convert the field.

Between the pines and oaks, whose seeds will arrive first? A large oak tree, making an admirable pile of acorns, sets off on a hopeful start to this race. She beckons a squirrel to come and carry one of the heavy payloads. The squirrel, inspecting several, chooses the tastiest-smelling nut. She grasps it in her jaws and hops daintily through the forest away from the tree. The tree beams as she watches the squirrel dutifully burying the acorn. A blue jay arrives as well, flying another acorn even further away before finding a clever hiding spot. The oak has her eye on that sunny field, waiting for some of her seeds to be carried there. Grinning mischievously while crossing the fingers on one of her branches, the oak now prays that the squirrel and blue jay forget where they placed the acorns, so her children become saplings instead of supper. Day after day, the oak watches the laborious work of the animals.

A pine glances over at the oak, rolls her eyes, and slaps her forehead. "Poor fool." In one fell swoop, the pine scatters thousands of winged seeds into the wind and across the field. And that's that. The pines have it. After a few years the farmer's field is awash with pine seedlings.

Basking in the sun, the pines grow furiously. Up and up they climb. The grasses and dandelions are soon completely shaded out. It's too dark for anything to grow beneath the pines. In the winter, snowshoe hares and grouse cut between the dense woody stems, munching on needles and hoping not to run into a bobcat. After decades of crowding each other, fewer and fewer pines remain. But the ones that do are big. Really big. Now they're full trees, eighty feet tall and ten to twenty feet apart.

We're standing in a forest. A few feet away a small mound of orange pine needles holds the scent of the healthy bobcat that made it. Some light filters down to the forest floor through the canopy needles, but it's still pretty dark. Certainly too dark for offspring of the sun-loving pine trees to survive. Pine saplings crave light and cannot survive beneath a canopy of their own species. But wait, what's this? This little set of dark green leaves, broadly lobed and tinged with red, gripping a tiny stick in the ground? There seems to be a little oak seedling emerging from a small hole. In fact, I see many of them. Who put these oaks here? They appear to be surviving just fine in the shade. Not growing furiously, but hanging on. Slowly, year after year, the little squirrel-planted oaks stake out their place in the forest understory and wait. And wait.

At last the hasty pines reach the end of their lifespan and begin to topple. The patient oaks are ready. They've waited a long time for this moment, and taking their cue, they spring toward the open sun.

The farmer never made it back to her land to see the forest that replaced her field. But many generations later one of her descendants, tired of digging irrigation ditches in Colorado, at last decides to go back East and search out the family roots. There, near the remnants of a stone cellar, the ditchdigger finds a magnificent oak forest on the old family farm. Sturdy oaks, patient and steady, won the race. Or did they?

Wandering beneath the oak canopy, the ditchdigger notices very few oaks in the understory. That the understory is different from the canopy tells the ditchdigger that the forest is still in flux. It's such a simple pattern to look for, and one we rarely do. I grew up in a Tennessee forest where giant tulip poplars stood watch over an understory deplete of tulip poplars. There beech and sugar maple promised to succeed into the next generation of canopy trees—the forest was in transition. What I saw growing up as an immutable old forest was itself merely growing up as well. Now I see tulip poplar forests in the South the same way I see white pine forests in the North—echoes of the old farm onto which these wind-dispersed, early successional species began the forest (fig. 3.6).

In the forest the ditchdigger inherited, he sees mostly beech saplings. With nuts flown in via blue jay, these beeches are even more patient and shade-tolerant than the oaks. More than patient, the beeches bring some secret weapons. Beech saplings can grow in the dark understory, adults cast a very deep shade on the competition, and they can reproduce clonally from their roots. When at last the oaks die and the beeches take over, this ultimate of long-lived hardwoods (at least in our region) may even employ chemical warfare. That is, they lace the soil with toxic compounds that kill other trees, so that only beech saplings survive beneath the beech

Figure 3.6. Early successional forests: (a) young white pine forest, (b) white pine forest with hardwood understory, (c) red pine forest with hardwood understory, and (d) tulip poplar forest with beech understory.

forest—a tactic called allelopathy. Only then, with the climax beech community, will the forest have reached a stable equilibrium.

At least that's how the story goes. This is a simplified version of forest succession, admittedly an imperfect fit for Maggie's Forest. For one thing, the eastern coyote we imagined prowling the meadow of a prior century hadn't even been invented yet, although that's a story for another site. More to the point, beech prefers moister soils than these, and which late-successional species—beech, hemlock, sugar maple, or someone else—dominate the climax community all depends on the local site conditions. But this narrative also sup-

poses that "stable equilibrium" is even a meaningful concept. Few ecologists really believe that any more.

PRIMARY SUCCESSION

The forest succession in the farmer's field was really *secondary* succession—it wasn't the first time plants had grown there. The ecosystem was given a jump-start from all that plowed soil, rich with organic material. *Primary* succession is when the system

Figure 3.7. Primary succession at Indiana Dunes: (a) young dunes with beach grass in foreground and lone cottonwood in background; (b) middle-aged dunes with junipers, fragrant sumac, and other shrubs in foreground and young trees in background; and (c) old dunes with a mature oak-maple forest.

starts from scratch, like bare rock left behind by a retreating glacier, newly formed sand dunes, or Charley's favorite abandoned parking lot—where tiny colonizers have been building soil on top of asphalt for the past few decades.

On the last night of a forty-day, 15,000-mile road trip, Charley and I rolled into the Indiana Dunes National Lakeshore after midnight. This was one of the first places that primary succession was formally charted out by ecologists. Charley fell asleep in the passenger seat, and I wandered out to sleep on the beach. In the morning I awoke to the crashing sounds of Lake Michigan, a cold gray wind, and the sight of a nuclear cooling tower looming disproportionately large over the lake. Charley soon came racing down the beach, agitated. He had recently awoken, noticed me missing from the driver's seat, and saw the furious waves in front of the car. He assumed I had drowned in the lake. After assuring him I was still alive, we wandered inland over the dunes, retracing the steps that ecologist Henry Cowles had taken a century prior when he made this place famous.

With my sleeping bag near the shore, I had slept on pure sand. Shifting, dry, and lacking organic matter, this is not a friendly place for plants. Walking away from the water, Charley and I began to encounter low beachgrass scattered across the dunes (fig. 3.7a). This pioneer species is the first to tame the hostile sands. Because the lake levels have dropped historically, the farther we walked from the receding shoreline, the older the dunes. As we walked, the species shifted. A bit further from the shore, juniper and a few other shrubs came to dominate (fig. 3.7b). As the dunes aged, larger pines replaced the junipers. Finally, forests of oaks and maples arrived, along with the whole rich suite of familiar plants that live in typical eastern forests (fig. 3.7c).

Each stage in the Indiana Dunes seems to depend on the prior stage. The pioneer species stabilize the loose sand of the dunes, injecting the first layers of organic matter. The developing soil now holds just enough nitrogen and moisture to allow junipers to grow, which then shade out the pioneer species. At first there's still not enough soil to support the bigger trees. But after the junipers have worked for years developing the soil further, the pines begin to enter and prepare conditions for the final wave of oaks and maples, which by now have waited hundreds of years for this moment.

Isn't it lovely how plants at each time step set the conditions to help out the next wave of plants? The pieces of nature are so well ordered and cooperative. The system grows like a well-designed "super organism" as it develops toward its predestined climax. At least that's what the esteemed ecologist Frederick Clements would have said. But he's been dead a long time.

There was another ecologist, Henry Gleason, who famously challenged Clements's view of nature. Are all of the tree species really working together as one well-designed system? Or is that just our rosy view of the world? Maybe, deep down, each species is just selfishly doing its own thing competing for space. We might like to think that each species in the Indiana Dunes is an intricate part of a coherently evolved natural community; take away the oaks or the pines and the whole thing would deteriorate.

But consider the evidence that Margaret Davis brought, digging down into pond muck to find prehistoric pollen grains from thousands of years ago. With these microscopic fossils, she created maps of where trees lived in the past, reconstructing the march of species over time (fig. 3.8). Back when glaciers covered most of North America 20,000 years ago, neither oak, maple, beech, pine, nor hemlock would have been found on the shores of Lake Michigan. No, there was just a mile-high ice sheet there. The pines and hemlock, according to Davis's pollen maps, were confined to a small region on the mid-Atlantic coast. At the time, the oaks, maples, and beeches were all down in the south—with oak extending farther east than maple and beech. Their ranges barely overlapped in space, so how could these species all depend on each other? Only later, after the ice sheets receded and each species migrated in its own peculiar path northward, did they end up in the same place. The maps have been updated since Davis worked on them, but the basic idea remains the same: over glacial time frames, species move around independent of each other, and which species grow together is constantly in flux. When you start taking such long views of the world, the idea of perfectly balanced, stable natural communities begins to crumble.

This debate—about whether the trees are working together for

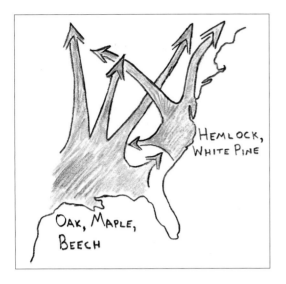

Figure 3.8.
Generalization of tree species migrations over the past 20,000 years. Following Davis 1981.

the good of the whole or acting as self-interested individuals—really cuts to the core of your philosophical outlook on life. Instead of a sugar-coated lullaby, for the last four years, my standard bedtime song for the kids is Neil Young's "After the Goldrush," often followed by "Don't Let It Bring You Down," "Ohio," or "Needle and the Damage Done." Strumming his ukulele, Juno belts out these songs with more understanding of their dystopian lyrics than you'd expect from a four-year-old. A few weeks ago I saw that Neil Young was on a solo tour, but the only tickets left for the nearby show were being hawked for over $1,000. However, there were still some reasonably priced seats out in Detroit, 700 miles west. From there it's only 250 miles more to the Indiana Dunes, which I was dying to get back to. So we all got in the car with our ukuleles and headed west.

The day before the concert, I arrived back at the Indiana Dunes for another look, this time with the sort of distrustful lens that might come from listening to too many Neil Young songs. Maybe it's not that the pines and oaks are waiting for the beachgrass and other pioneers to set the stage for them, but maybe it's more a matter of random probability. Pine seeds are eaten readily by mice and, to arrive at the beachfront, must travel a fair distance from the adult trees. Given this, it's a rare event that pine seeds end up in a position to sprout. The seeds of beachgrass, on the other hand, are abundant on the young dunes, disperse easily, and aren't as prized by mice. So by sheer chance more beachgrass seeds end up sprouting in the unsettled sands. Plus beachgrass spreads clonally, so it can often skip the whole seed stage altogether. Also, pines grow much more slowly and only sprout in wet years. So even if they sprouted side by side, the slow and steady pines don't end up growing very big until after many years in which beachgrass has been partying in the sand.

Rather than beachgrass helping pines in an orderly procession, it's just each species for herself in a random process. It turns out that the nutrients in the soil have basically no effect on the survival of species in the successional stages of dunes. Rather, soil development is just a by-product of the process. In a big experiment mixing up seeds and saplings of many species all across the developing dunes, that's exactly the conclusion that modern ecologist John Lichter came to.

CONSERVING SUCCESSION

Whether succession is random or guided by destiny, those species at the early stages of succession can't survive under the shade of the late-successional trees. If you want early successional species in your forest, you need to either pray for some natural disturbance to knock the trees down or cut them down yourself. And so a lot of species management in eastern forests involves cutting trees down and keeping them down. Consider the bobolink, a bird that is the focus of many conservation efforts in the East. This bird needs large fields, and so to keep it around here, we must keep mowing the fields to prevent trees from growing.

But why is a bird so dependent on lawnmowers? Out in the Midwest, natural forces—like a drier climate combined with large herbivores and occasional fires—can keep the landscape as prairie without the intervention of mowers. So bobolinks are much more common in the Midwest. It's just that in the East they need help to survive. Which raises the question: Do bobolinks even belong in the East? Why waste time and money supporting a bird that doesn't even seem to belong here?

In the mid-1800s, grassland birds like bobolinks were much more abundant in the Northeast than they are today because most of the landscape was covered with farmland. Perhaps they were historically limited to the western prairies, but when humans created artificial prairies in the East by cutting down all the trees, the prairie-dependent species moved into this new habitat. Then farmers abandoned their fields and forests regrew. Eastern populations of grassland birds have been declining ever since. With this view, bobolinks are just relics of an artificial prairie that maybe never belonged here in the first place. If we just removed humans from the equation and let nature take its course, recovering forests would probably wipe out the bobolink habitat. So maybe nature is telling us to forget about the bobolinks here. Although, arguably, there's nowhere that the bobolinks are safe, since prairies in the Midwest are facing increasing threats for other reasons.

Rather than trying to restore the artificial landscape of the 1800s, shouldn't conservation aim for something more pristine? Like, maybe we should be aiming for the landscape prior to 1492, prior to the massive destruction and change brought by Europeans? But then again, Native Americans had a heavy impact on many parts of the North American landscape as well. They farmed. They set fires to clear undergrowth for hunting and to encourage blueberries. Indeed, they may have maintained bobolink habitat in parts of the East as well. So if we're looking for a "natural" reference point free of human meddling, maybe 1491 isn't great.

OK, so what if we roll the clock all the way back to before Native Americans arrived? After all, when humans first invaded North America about 13,000 years ago, it marked a major turning point in the local nature. We helped drive to extinction North American mammoths, lions, cheetahs, camels, horses, and other large mammals. Does our moral obligation to cleaning up after our own species require us to look back that far? Some conservationists think so and propose "re-wilding" North America by introducing African and Eurasian elephants, lions, cheetahs, camels, and horses to our continent to restore natural processes and as a way to preserve these globally endangered lineages. Seriously.

What happens if we take that long-term perspective with bobolinks? Roll the clock back to 20,000 years ago. The glaciers were just receding, leaving in their wake a large swath of savannahs and grassland extending from the Great Plains all the way to New England. As millennia rolled on, the forests regrew and eventually dominated the local landscape for most of the past 8,000 years. But pockets of grassland likely persisted. At first mastodons, caribou, and peccaries trampled out open spaces. After these creatures left our local landscape, the still-abundant beavers created and abandoned ponds that formed meadows. Windstorms battered vegetation that on poor soils didn't regrow quickly. And people planted crops and set fires. From Georgia to Maine, many areas that are now within the eastern forest were, at times in the past several thousand years, covered in grass. Bobolink habitat.

Wherever we look in the past, it seems there were grasslands in the East. Yet, left to her own devices, nature seems intent on a future with very few grasslands here. Perhaps the distinction between the "natural" world and the human-modified world isn't helpful. For a variety of reasons—glaciers, climate, humans—bobolinks *are* in the East and have been for some time. The question is, do we value having them here? It's the edge of their range, so maybe we should just focus on protecting the birds in the core of the range. But sometimes the most important part of a species' range is at the edge. That's where evolution is pushing and pulling the most, and individuals at the edge might determine the future plight of the species. And these bobolinks in the East may be an important component for saving the whole species, which, like most grassland birds in the United States, has been dramatically declining for the past half century.

For whatever reasons, the people here like bobolinks and have decided to manage for them. To do so requires an active approach, with chainsaws, mowers, tractors, and matches. Of course, it's not all just for bobolinks—we're creating habitat for many similar early successional species while harvesting the agricultural and aesthetic resources that people draw from managed land.

The same logic explains the intensive management at Maggie's Forest. In the surrounding wildlife management area, we see a patchwork of pitch pine forests, scarlet oak forests, scrub oak thickets, and grasslands. We find barrens buckmoths, William's tiger moths, New Jersey tea inchworms, and wild lupine—all species that are rare in this region but common elsewhere. We find barrens metarranthis and spreading tick trefoil—species that are globally rare. Indeed the entire barrens community is globally rare. Today they seem to all depend on human intervention. The fire-adapted species are here because humans have been burning this pine barrens for at least 2,000 years. Without continued burning and cutting, white pine would replace pitch pine and closed-canopy forests would replace the thickets and fields. And in that transition the rare species dependent on this unique community would be lost.

That's not to say that fires are always a good thing in every ecosystem, nor are all wildfires the same. Some fires—surface fires—just creep along casually through the leaf litter and low shrubs. The ground-level heat might kill the trees, but often these surface fires do little more to trees than simply paint the bases black. But fires can also dramatically alter a landscape, especially in places not used to fire. Sometimes, for instance when people burn the trees on top of a mountain, they can cause all of the soils to wash away—soils that may have taken thousands of years to develop—creating a barren mountaintop and resetting primary succession.

Down in West Virginia there's a rare little salamander, called the Cheat Mountain salamander, which lives in the thick, moist layers of litter in high elevation red spruce forests. In the late 1800s and early 1900s, West Virginia's red spruce was extensively logged, leaving the forest floor exposed to the harsh sunlight. The layers of spruce needles, mosses, and lichens all dried out and became the fuel for massive fires. Deep, subsurface fires burned for months at a time, down to the bedrock. In 1914, for instance, a fire in one canyon started in May and burned until it was put out by November snows. Huge boulders that were once hidden underground rose to the surface as the soils burned up and washed away.

These fires and the loss of soils were terrible news for the Cheat Mountain salamanders. Luckily, throughout the fires, there were a few scattered boulders that held on to moist pockets of soil beneath them. It was in these spots that the salamanders clung to life during the blazing fires of the last century. And so today most of the surviving Cheat Mountain salamanders live right around the boulders that saved their ancestors.

Of course, species conservation is only part of the equation. Here at Maggie's Forest, fire control—and the associated risk to lives and property—is the highest management priority. Low, controlled ground fires are OK, but not raging crown fires leaping from tree to tree and then possibly to nearby buildings. Pitch pines, full of flammable oils, are highly susceptible to such fires, especially when the trees are packed tightly together. So how do you prevent such spreading crown fires? Thin the trees out so much that the canopy isn't closed and the trunks are widely spaced—cre-

ate a savanna. Then light many low fires to keep the forests from closing in.

This natural system is maintained through cutting and burning. And it has been for a long time. Just take a look at the oaks in the chapter-opening image: multiple-trunk trees. There's one at the center of the image and one way in the background. And there are many more off the frame (see fig. 2.3), evidently from logging in the recent past. In fact, in 1939 less than 1 percent of this pine barrens contained closed-canopy forest. A hundred years ago this was a place for growing corn, cutting timber, and extracting tar and turpentine from the pitch pines. Now the goal is to preserve a dynamic pine barrens ecosystem (fig. 3.9).

Figure 3.9. Fresh scars of intensive management adjacent to our field site.

But why *here*? How did the pine barrens get here? Because this place is special. At Bark Hollow the underlying bedrock was different from the bedrock in much of our area, and this created a natural community that was correspondingly unlike its surrounding communities. Maggie's Forest also sticks out like a sore thumb. If we were on the coast or in New Jersey, it'd be one thing. But a pine barrens here, far inland, surrounded by forests of maples, oaks, and white pines for hundreds of miles?

When we arrive here, one of the first things I ask the students to do is to collect some big rocks and bring them back to the group so that we can figure out the underlying geology. They dutifully search and search, dig and dig, and inevitably come back with the biggest rocks they can find: about the size of a blueberry.

"You can't do any better than that? Go out and find me a nice big rock that we can actually work with!"

A while later they come back, heads hung in shame with a rock the size of a walnut.

I ask, "What's the deal?"

At last they declare that "there just aren't any rocks here!"

And that's the truth. More or less. Dig as you will, you mostly just hit the same fine-grained sand and gravel, with some occasional bigger stones mixed in. And it's easy to dig here. That's why we bring along the soil corers, those metal tools in the bottom of the chapter-opening image. Jab them into the ground, pull them out, and voilà, a study of the soil. Try that back at Bark Hollow, and all you'd get is the clanking sound of metal hitting rock and a little dent in the tip of the empty instrument. But here the soil core captures a beautiful image of natural soil layers. The top few inches, rich in organic material, are darkly stained with decaying needles and beetle poop. A few inches down, the soil is bright and clean.

And the sand just keeps going. In fact, I think you'd have to dig sixty feet or so before you hit something different.

So this whole place is a giant sand pile. Well, maybe not quite a pile. I mean, when I think of a pile, I think of something shaped like a cone, like a mountain of sand. But this place is perfectly flat. Flat, flat, flat. Still, why aren't there any bigger rocks here?

In geologic terms the sediments here have been well sorted. We don't see different-sized particles all jumbled up together. Rather, we see that particles of the same size have all been sorted into appropriate piles. This tells us that the particles have been carried, by either wind or water, some distance from their original source.

Consider a small river carrying a bunch of variously sized rocks. Find a bend in the river, stand on the inside bank, and look across the river to the muddy cliff on the outer bank of the curve. Take your shoes off and wade across. As you first step in, notice that the water is moving slowly. But as you wade deeper and deeper, you have to fight harder and harder against the current to stand up. By the time you reach the steep wall of the outer bank, you are leaning considerably against the river's force.

Water sorts particles because the flow of water is not uniform. The fastest parts of the river can move boulders, but water in the slowest parts can barely lift a sand grain. So sand grains only settle to the bottom in the slow parts of the river, like the inner bank. The river would never let a sand grain rest in that fast current by the outer bank.

Even smaller than the sand grain are tiny particles of clay. It takes barely a swirl of motion to keep clay suspended in water. And it's not all about size.

Remember from Bark Hollow the end fate of rocks exposed to weathering on the Earth's surface? The two primary products of weathering are fine-grained quartz and clay. Although some people use "clay" to refer to any type of very small particles in soil,

I use "clay" in the sense I learned in geology class—as reference to a particular chemical composition. When defined from this chemical perspective, quartz and clay are silicate minerals. They are primarily composed of silicate tetrahedrons—that is, a little pyramid with four oxygen atoms at the corners and one silicon atom in the center. In quartz all these little pyramids are stacked up in a three-dimensional network. As big quartz rocks weather, they break down in three dimensions, maintaining a more or less rounded shape. However, in clay the silicate pyramids are tied together in flat two-dimensional sheets. When clay minerals break down (and there are different types of clay minerals), they split apart in layers. Like separating a ream of paper into individual sheets, each tiny piece of clay has a large surface area relative to its volume.

In the water column a clay particle hangs, responding to every push and shove from the water just like a sheet of paper. Compared to clay, quartz sand grains, round and solid, drop like bowling balls through the water. And the relationship between water and soil particles cuts both ways. Just as a column of water can't hold on to bowling balls for very long, a pile of bowling balls doesn't do a good job of holding water particles. Pour water on a big pile of sand, and it will fairly quickly drain through to the bottom, leaving the top sand relatively dry. But a deposit of compacted clay, with its interwoven reams of paper tightly adhered to each other, is loath to let a drop of water pass through.

In our river a sand grain and a clay particle are carried along together in the swift current. Suddenly, the river meets a lake. The water slows down. The current can no longer lift the little bowling ball, so it falls out of the water column along with all the other sand. But there's still enough current to carry the sheet of paper, which won't settle down except in the stillest of waters where all the last little swirls have exhausted themselves. The clay drifts far out into the middle of the lake. There, finally, the clay particle gently saunters down to the bottom.

THE LAKE

So did you find a source of flatness in nature? In your cup of pine needle tea, perhaps? Or out in the canoe? Water. I can't think of anything flatter than the surface of water. That's what we're looking at.

We are sitting on a delta. This is where a flowing river suddenly meets standing water and dumps its sediment (fig. 3.10). The land around New Orleans grows ever further into the Gulf of Mexico because the Mississippi River brings fresh sediments year after year. Before inching forward, the river piles the sediments as high as she can—right up to sea level. Or, in our case, lake level. The land is flat because it's the surface of a lake.

A glacial lake. The lake that 15,000 years ago lapped up against Bark Hollow's basalt mountain. Full of meltwaters from the retreating glacier. Dammed up by a giant pile of debris that the glacier dropped into the rift valley farther south. Two hundred miles long and, in places, over 300 feet deep.

It only lasted a few thousand years, but that was enough for substantial deltas to form from Massachusetts to northern Vermont, like the one we're sitting on. About the same time, the first people were arriving here. I'm not sure there's any hard evidence that people saw the lake, but Native American oral histories still feature an ancient giant lake that once filled the valley. Can such information really be transmitted over 500 generations of the game of telephone? I like to think so. Somehow it gives me comfort to believe that people once set eyes on this lake, admiring its expanse and swimming with the mammoths.

The lake is long gone now. All we can do is read about it in the

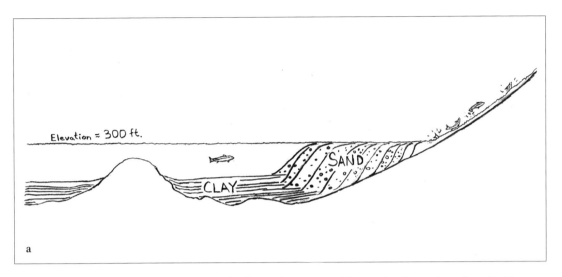

Figure 3.10. (a) When a river enters a lake, sand and gravel are deposited in a flat-topped delta near the shore, while clay drifts out and settles farther out. The delta height of 300 feet represents modern elevation at our field site. However, the actual elevation 10,000 years ago would have been closer to sea level; ever since the weight of the glaciers melted away, the land has been rebounding upward. (b) Another delta near Maggie's Forest has been turned into a gravel quarry.

pages of the land. The dry-loving pine barrens describe soils that are deep and sandy—an open drain for water. The horizon spells out the lake surface. The slope we climbed to get here—the underwater face of the delta—documents the lake depth. The fish hatchery on the road at the base tells us of the abundant springs from all the water that slips down through the sand, only to hit the clay layer below and be shunted sideways.

To be clear, not all flat, sandy pine barrens are ancient deltas. Travel across New Jersey and you will see other features that tend to be flat, sandy, and pitch-piney. As the glaciers retreated from the coast, they often left behind broad, flat landscapes of sand and gravel smoothed by myriad braided streams carrying meltwater—

called the glacial outwash plain. It's the context here at Maggie's Forest—a discrete pine barrens on a flat terrace with a steep slope down to "regular" woods in the valley below—that point the finger to a lake delta.

OAKS AND PINES

Our tea is down to the last drips, but we still haven't quite answered the original question: Why are the trees different on the left and right? If we were to poke around on the ground beneath the trees, the low understory species follow the same pattern.

Fire? Wind? Disease? Looking at the aerial photo in figure 3.2, none of these natural forces would seem likely candidates for driving the oak versus pine pattern. In this winter image the pines, still holding their needles, are visible as dark green. The bare oaks are light brown. The borders between these dark and light patches are straight. Rather than the ragged, irregular shapes that nature would opt for, I see rectangles.

Right angles and straight lines. This must be the work of humans. Not the recent intervention, but the remnants of land use from one hundred years ago. Back when Maggie's Forest was a farm. The giveaway is in the soil cores on the bottom of the chapter-opening image. You see, only one core shows a distinct A horizon. That narrow black band at the top. In the other core the top eight or so inches of soil have been mixed into a uniform blur. The farmer's signature, written with a plow. One side was plowed for farming. The other side? Maintained as a woodlot for logging timber—so say the multiple-trunk oaks.

One hundred years ago both patches might have looked similar—wide open and devoid of forest. An empty field, freshly plowed. A sea of stumps, the last of the trees cut down. The farmer flees. Then what? In the field, winged pine seeds outcompete the clumsy acorns to win the first heat of the race to forest succession. But in the old woodlot the oaks have the advantage. The logging operation only removed half of every tree. Underground, huge root structures, full of stored sugar, ready to capture every particle of passing water, sit poised for spring. The tiny pine seeds don't stand a chance against such massive reserves. In the cut woodlot the oaks leap to the canopy, sending multiple shoots from every stump, shading out the pines in the blink of an eye. And here we are today, a hundred years into the first wave of forest succession. One hundred years on, the trees are still just a shadow, outlining the farmer's saw and plow. What legacy will today's cutting leave?

- Consider the shape of the land. What surficial geologic processes are responsible for having moved, sorted, and shaped the sediment?
- What are the successional dynamics in your region, and what successional stages are represented in your site? Look for evidence of slow species turnover.

a

197 ft

a

b

b

Figure 4.1. Know your place in the
broader landscape. (Inset map: USGS)

4. Context

Mid-July, ten miles southwest of Maggie's Forest. I have only a couple hours and a lot of walking to do—this site has two parts. I ditch the car on the side of the highway, throw my backpack on, and cut into the woods. Dressed in a T-shirt, I think to myself, "Is there some reason I need long sleeves? No, it's a warm day." As I set off alone, I'm propelled by the memory of my students' bubbly anticipation, still hanging in the air from years ago. I warned them that this would be a long day and they needed to prepare for wet and muddy feet. They took the warning as a promise of adventure.

I start along an old logging road and find a collapsed bridge over a straight-walled creek. Crossing the creek, I forget about the trail and wind my way westward. It's fairly flat here, just scattered hummocks, bumps, and holes. Weaving alternately through patches of spicebush and mountain laurel, in about twenty easy minutes I make it to the first spot (fig. 4.1). Towering tupelos, pin oaks, and swamp white oaks provide shade for royal ferns, highbush blueberries, sedges, and mosses.

I set my gear down and start searching for the best picture angle. Swarms of mosquitos keep me moving quickly, hands flapping, as I crane my neck to read the canopy trees above and peer left and right through my lens. The topography is so uniform that, in my funny little dance, I lose my orientation. A bit panicked, I spin around and find a sense of relief when I recognize the solid trunk of one of the towering oaks.

In front of the oak is a little flat spot devoid of vegetation in which I see the students still digging a hole trying to solve the site's puzzles. They pass first through darkly stained bits of leaves and muck and then into layer after layer of grayish silt and clay. Four years on, they're still digging down through the clay. Other students are out collecting leaves, testing soils, lifting logs. At last they all finish their observations and come together in a circle between the oak and a tupelo.

Josh stands quietly, tucked discretely under the oak's arm, with his head a bit tilted as he listens to the other students. Calling out species names and pointing to the maps, the students spiral around ideas about the history that formed this site. Exhausted and a bit defeated, the students' eyes narrow as the ends of their theories don't meet. Josh gently steps forward and, gesturing with his hands, pulls together the whole story. Light returns to everyone's eyes.

It all makes sense as I plod through these apparitions back to the old oak, expecting my bag, tripod, GPS, and compass to be piled at the base. But there's nothing there. This isn't the same tree at all. My gear has drifted even further away. I need to slow down and pay more attention.

I stand motionless. I breathe. I center myself, relaxing back into my spine. I look in all directions, using my peripheral vision as much as my forward tunnel vision. When I start to move again, my body moves slowly and with intention. One foot goes up, tracks forward, and eases down—the whole while I am balanced on the planted foot. If needed, I could freeze at any point in the stride. In this way I find my gear again. I breathe relief. I snap some photos. Then, falling back into a faster rhythm, I trudge on to the next stop.

As the land starts to tilt upward, a little wood frog hops in front of me. I spring down to catch a picture. How many times have I gone crashing through woods in early spring destined for a lit-

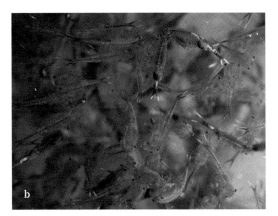

Figure 4.2. Vernal pool specialists: (a) mating wood frogs surrounded by egg masses, and (b) fairy shrimp.

tle cold pond, plunged my hand into the icy water, and picked up a wood frog's translucent jelly-like egg mass? At the 560 ponds I visited for my dissertation, I counted nearly 10,000 wood frog egg masses. Still, every single one brings me a thrill.

Wood frogs are vernal pool specialists (fig. 4.2). In early spring, on the first rainy night, they leave their winter hideout in the forest and hop down to the pond, where the ice is just beginning to thaw. Males call out a cacophony, like a sea of laughing ducks, while a plump reddish female swims about looking for a mate. She lays her eggs as a grape-sized sphere with hundreds of little dark beads. Within a few hours the egg mass expands to the size of a grapefruit. Within a few weeks the larvae hatch out and begin eating algae. Within a few months the larvae transform into froglets and hop away just as the pond is about to dry out. It's important that the pond dry out—that kills off the fish, large dragonfly nymphs, and other long-lived predators that might otherwise lurk in the water and eat the defenseless tadpoles.

After I'm satisfied by the frog in my lens, I glance toward my elbows sinking into the wet mud and notice delicate toothed leaves kissing my arms. Poison ivy. If only I'd worn long sleeves. Luckily, there is a little puddle nearby—possibly the remnants of the vernal pool that my little friend hatched out of. I go through my poison ivy ritual: spread silky mud onto the affected area, scrub with dirt, rinse with water, repeat. Some folks swear by jewelweed, some swear by specialty soaps, but I think immediate application of the closest grit and water works wonders on poison ivy, washing away those itch-causing oils. Hopeful, I continue the gentle climb to the next spot.

Soon I'm there. The trees are entirely different: an upper canopy of mostly northern red oaks with an understory dominated by beech and a few small sugar maples and hemlocks. I recognize a heavy gray rock with rounded edges, the size of a toaster oven, under which my students had found a redback salamander. No salamander today. Instead, a flurry of tiny reddish ants scramble to cart their white larvae to a deeper spot safe from rock-lifting hominids. Meanwhile the air has warmed a bit, and the mosquitos are even more active. I unfold my tripod, stretch out its legs, and set its feet in the leaf litter so that it stands shoulder to shoulder with me. I wander over to a beech tree to have a look at the scars on its bark. Walking back toward my tripod, I rub my eyes trying to figure out why everything seems blurry. A little fuzzy dark cloud is hovering around the tripod. A swarm of mosquitos, dripping off the tripod head in opaque masses. Hundreds and hundreds of them. I've never seen anything like it. I really wish I'd worn long sleeves.

I take pictures as fast as I can and head for safety. About a hundred feet west, a long pile of boulders—an old stone wall—marks the boundary between the forest and an open field where courting bobolinks abound in the spring. The mosquitos don't dare follow me into the field, where the air is too dry for them today.

On our topographic map, the field shows up as white, whereas the forest is light green. As a military outgrowth, such maps historically used green to indicate "vegetation with military significance, such as woods, orchards, and vineyards," according to army field manuals. The modern makers of these topo maps probably aren't thinking in these terms, but this does make you wonder. How, why, and when were these maps created? Most of the topo maps we use around here were published by the USGS in the early 1960s. How did they know where every hill, valley, creek, forest, and field was? Because people systematically walked the landscape, carrying surveyor rods, compasses, pen, and paper. An enormous undertaking. Wherever I go on the map, at least one person—a tired government employee carrying a tall pole with measuring marks atop it for his partner to read from a distant vantage point—has stood here before. Since 1884, the USGS has been sending dedicated crews out onto the landscape. Actually, back in 1777, before the United States was even formed, General George Washington appointed a geographer to the task of making maps. They now cover more than three million square miles. When I speak of topo maps, I'm smelling the dry, papery, slightly musty smell of the closet where my dad kept them rolled up in a poster tube when I was a kid. My eyes widen with thrill when he opens the closet, the light flicks on, and the brown tube adorned with funny old stamps descends from the top shelf. Out of the tube comes a long, slightly yellowed, cylinder of paper. Unrolling them on the floor, he traces his finger along the ups and downs of hills around our house. Although modern technology is rapidly replacing such paper copies, many GPS devices still rely on the 1960s topo maps, and these maps were created through a long and hard-fought field collection effort. On the flip side, those data are now over fifty years old, and things change— so you should always take every map you use with a grain of salt.

FOREST CONTRASTS

The central question at this site is: Why are the trees so different in the two places?

In a broad sense the trees aren't so different. The forests here, like at Bark Hollow and at Maggie's Forest (prior to cutting), feel generally like home to me. I'm a creature of the temperate deciduous forest biome, a creature that rests under the safe, dim blanket of maples, oaks, pines, and beeches—despite the mosquitos.

The uphill site could be almost anywhere within the eastern forest of the United States. Sugar maple and beech are two species typical of the northern hardwood forest type, which dominates New England landscapes a bit north of here before the transition to boreal forests of the far north. Northern red oak, on the other hand, is more typical of the oak-hickory forests to the south. At first I tell myself that the overlap of these three species here pinpoints our location to southern New England, in the transition zone between these forest types. But then again, in the hills behind my childhood house in Tennessee, there are places with those same three species that look almost identical. Except for the rocks, I suppose.

The downhill site, however, has more unique character. Swamp white oak, pin oak, black tupelo (aka black gum), royal fern. These species bring me to very precise locations: tupelos guarding a pond filled with Jefferson salamanders in the middle of a hemlock forest; pin oaks scattered in a dried-up vernal pool in a red maple forest on the edge of a cornfield; a grand swamp white oak in a muddy wetland surrounded by an upland of red oaks; royal fern gracing the edges of a stream cutting through a larger forested wetland, which in turn is encircled by a northern hardwood forest. In all the miles of forest I hiked through on my way to the hundreds of vernal pools in my surveys, I don't think I passed a single swamp white oak, pin oak, tupelo, or royal fern, except at intervening wetlands I accidentally encountered. All are wetland species. Because this wetland is dominated by woody plants—trees and shrubs—it's not a marsh. This is a swamp, with indicator species that relay a strongly coherent message. In fact, this place is called the "Great Swamp."

If you're from Pennsylvania, Tennessee, or anywhere in between, perhaps you'd complain about my list of wetland trees; you might think one of those trees doesn't belong. Black tupelo. Down south I find tupelo abundant on hilltop forests right next to chestnut oak, in dry, nutrient-poor soils. In New England I've never seen tupelo except in a wetland. Certainly not next to a chestnut oak. These seem like pretty opposite habitat types. What's going on? I honestly don't know.

It's useful to remember that species behave differently in different regions. Maybe in warmer climates black tupelos like dry sites, whereas in cooler climates, they like it wet. In Massachusetts I grow my beard out, in Tennessee I shave it off, in Arizona I sport a Fu Manchu. Although, I wouldn't be surprised if it turned out these upland and lowland black tupelos were secretly different species—for decades scientists have been debating how to lump or split the various tupelo species based on morphology, habitat, and genetics. All I can say is that here, this tupelo advertises a swamp.

You might also be familiar with pin oak as a planted tree on college campuses and other landscaped areas—areas that aren't wetlands. In some cases the land might have actually been a wetland before it was converted to a college, and the tree is just hanging on. But usually the pin oak was just planted there because people like them.

The hardest time for a tree is when it's very young. From the time the seed sprouts through its sapling years, a forest tree faces horrendous challenges—find water, find light, compete with neighbors, ward off herbivores—with almost no stored reserves to draw on. Adult oaks may be mighty, but even a dainty little fern on the forest floor can doom the helpless oak seedling growing beneath it. Often it's the species' unique strategies during that critical youthful stage that determine what habitats it will survive in.

Once it's high in the canopy, the tree has stores of energy in its trunk and roots, it has plenty of access to sunlight, is beyond the reach of many herbivores, and is now the one that casts shade on the struggling saplings below. When a gardener has tenderly nursed the tree through its sapling years, she can then plant it out in an open lawn and the old tree can be found thriving far from its natural habitat. But when found in a natural forest, the pin oak, like the tupelo and swamp white oak, tells us there's a swamp here.

But why is the Great Swamp here? The answer has already been—obliquely—laid out in the previous chapters. So while you ponder that, let's dwell on the question itself.

It's a question that doesn't get asked enough: Why is it here? People get hung up on the "what" and forget about the "where." In this modern world our spatial senses have atrophied. But context matters. Why is it *here*? As the animal tracker Sue Morse would say, "Half of tracking is knowing where to look; the other half is looking there." Knowing what something looks like doesn't figure into this maxim. Location is everything. Why is it *here*? "It" could be anything. A tree, a rock, a nest, a poop.

A POOP

Down in Tennessee for Thanksgiving, Sydne, Alder, Juno, and I met up with my friend John Norris to explore his land. Walking along a system of trails wide enough for his pickup truck, he guided us through a brushy field burned last spring to create bobwhite quail habitat. We turned left at a line of trees and descended through rows of overripe soybeans to the edge of a creek lined with raccoon tracks, then doubled back and made our way toward a forest above the burned field.

As we walked, John taught us about quail management, agriculture, and the local ecosystem. At several points along the walk, he ducked off the trail to check on his various wildlife cameras. Bobcats, deer, opossums, and squirrels flashed across the digital screens. John knows his land well; he knows his species well; he is a great naturalist. We approached an intersection where one trail led up into the forest, with John a few steps ahead. Before he turned to check on the last camera, John pointed down and, aware of my affinity for scats and eager to improve his own knowledge, turned to me and asked, "Raccoon?"

As I searched for the right words to respond, he told us he had seen the scat days before and looked it up. Sifting through Google images of poop with various tones, twists, and dimensions, he narrowed in on raccoon. The blunt ends, dark color, everything seemed to fit. But before I looked down, I knew it wasn't a raccoon. Before I had even caught up to the place where John was standing, I had a species in mind.

John's mistake was that he looked at the scat. If you want to identify a scat, don't start by looking at it. The first question is not, "What does it look like?" Ask, "Why *here*?" Here we are, at a trail intersection with a field full of quail on one side and a forest on the other. Raccoon scat *here*? Impossible. To be sure, the raccoon who pressed her hands into the creekside mud has almost certainly been near this spot. And yes, she would have been physically capable of relieving herself here. But why would a raccoon poop here? This just isn't the proper place for a dignified raccoon to poop. She would no sooner poop in this intersection than you or I would. No, that raccoon placed her scat in her usual latrine at the base of a tree, on the uphill side, near the creek.

Animals mark their territories in characteristic ways, with deliberate intention. Gray squirrels bite roughly furrowed bark of prominent trees on the downward-leaning side, creating a multi-shaded stripe that can stretch twenty feet or more up from the ground. Red squirrels, on the other hand, bite smaller round patches at the base of conifers and along branches higher in the tree. Groundhogs bite the base of woody plants near their den entrance. Deer thrash their antlers against small hemlock saplings, moose against larger saplings. Bobcats pee on the vertical surface of one-foot-high decaying stumps when available, or if a lone pine tree stands out in a thicket, the cat may be attracted to the soft needles under the pine, scrape up a pile, and scent the top of it. Otters choose the needles under a prominent pine on a jut of land sticking into a lake. Dominant predators generally like to poop in prominence, each species with its own specific tastes: on rocks or stumps, near important food sources, near someone else's mark, or in the center of trails—especially where two trails intersect. When each citizen places her note where it belongs, we all know where to look for which messages.

In our annual winter tracking course, Charley and I lead students crawling through a blueberry swamp full of snowshoe hares to our lunch spot: a corner of land that the swamp wraps around. On a map, it's a prominent location. We sit in a circle, one student leaned against a paper birch, one against an old, mangled hemlock sapling, one near the stump of a broken tree. The students don't know it, but Charley and I are playing a game. It's the game we've played for eleven winters now. Who will be the first student to notice the antler rub on the hemlock, the bobcat pee on the stump, or the bear claw marks on the birch?

As for the poop on John's land: the size of the trail, the intersection, the proximity to prey, the placement near the middle of the trail all screamed coyote. Just to be sure, John and I poked at the scat and found apple peels, still with a tinge of red. Definitely coyote.

BEARS

Up on the hill above the Great Swamp, the beech trees display prominent scratches from bear claws. Though less eye-catching, some of the tupelos down in the swamp also exhibit bear scratches. Why?

Claw marks of black bears come in two flavors. One flavor signifies intentional marking behavior. A big white birch on a travel route—a perfect place for reaching up and swiping to paint a sign that is both visually attractive and filled with the poignant fragrance of bear foot glands. Or perhaps a big oak—the biggest tree in the forest—a bulletin board with decades' worth of bear notes tacked to its face. Maybe a red pine along a road, dripping with sap. Or, equally good, a telephone pole dripping with creosote. Or maybe a tupelo on a hill in the middle of the bears' favorite dining swamp that she wants to claim for herself and her cubs. Next to such claw marks, the bark scowls with furrowed growth around bite marks. After she clawed and bit the tree, she turned and rubbed her back on the bark, imparting the fullest of her scent. A few wiry black hairs, pinched in the crevices, waver when you breathe on them. Spin around on your heels and look behind you to find small trees bitten and mangled. We can't decode it, but these marks are packed with information.

The second flavor of bear scratching is incidental: a mere by-product of climbing. Why climb? Various reasons. A bear climbs to escape from danger. A mama bear drops off her cubs in a big white pine, what Sue Morse calls a "babysitter tree," while she wanders into the adjacent wetland to eat emerging skunk cabbage leaves and other spring greens. In another treetop a lazy bear lounges in the crook of a branch. Although bears occasionally sleep in crude nests atop a tree, it's not typically rest but food that the bears are after. Usually, the bear biologists will say, "bear nests" that we see in trees are the unintentional remnants of feeding—branches snap as the bear bends them inward to reach the nuts at the tips, then the bear absentmindedly discards the branches in a pile. When trees hold aloft nutritious treats like tupelo fruits, cherries, apples, acorns, ash seeds, or beechnuts, bears climb to dine.

Here is a forest of beechnuts next to a forest of tupelo berries. Both fruits ripen in time for a bear to enjoy a satisfying supper in early autumn. Come spring, swamp margins fill with tender skunk cabbage leaves, quivering as the bear tears through the unfurling patches, devours this first post-hibernation salad feast and excretes cylinders made entirely of processed skunk cabbage. It's now clear why there are lots of bears here. In fact, it'd be a surprise if there weren't signs of bear here. Somewhere around there will certainly be intentionally marked trees, individual and prominent. But the claws climbing up to the top of every one of the beeches? Feeding sign.

The beech nuts are a special treat. Bears love 'em, along with turkeys, deer, blue jays, myself, and many other animals. But beech nuts are disappearing. Why? Just look at that beech tree standing in our chapter-opening image. The left side shows the classic smooth bark of beech. It's that slate-like surface into which vandal children scratch hearts around initials to memorialize relationships the trees will certainly outlive. Unless beech bark disease takes over. The right side of our beech is lumpy and mangled. Beech scale, a little insect native to the Black Sea, was accidentally brought to eastern Canada around 1890 and has since been spreading south and west across the United States. Lacking in wings and males, the female scales clone themselves and depend on the wind to randomly blow them to other beech trees to feed on. On a tree, a scale pierces the smooth beech bark and sucks out the juices below.

But it's not the scale that the tree is most worried about. It's the fungus that blows into the little holes left by the scale. Once under the bark, the fungus spreads. The bark bubbles and cracks. Nutrient flows are cut off. The wood weakens. Other insects, followed by woodpeckers and more fungi, invade. Soon the tree is so weak that, in a gust of wind a healthy beech wouldn't blink an eye at, the diseased beech snaps in half. Even if the beech doesn't die, it hasn't the strength to produce the volume of tasty beech nuts the animals once enjoyed.

HISTORY

Beyond just beech bark disease, this forest is far from pristine. The red oaks atop the hill stand above an understory in which red

oaks are absent. Soon the canopy will be beech, assuming they survive the bark disease. This is a forest in transition, still rebounding from past cutting. When I look north, each tree along a particular line of sight has a big scar near the base. Missing bark exposes inner wood like bones glaring through stripped skin. Each scar faces to the left. Where those scars are aimed, about eight feet out, there's another imaginary line along which all the trees have matching scars looking right back at the first set of scars. In between these two lines must have been an old logging road (fig. 4.3).

After being cut, felled trees were hauled along this road, perhaps first by horses and then later by a diesel skidder driven by someone like our eighty-year-old neighbor, who I still see wielding her chainsaw and driving heavy machinery through her woods. The logs were dragged down the road, bouncing and rolling left and right. The standing trees, watching from the side of the road, winced each time one of their fallen comrades banged into them, chipping another piece of bark off of the ever-widening scars. The logs, carrying tons of carbon in their wood, left the forest and passed into the adjacent field through the gate of the stone wall, bouncing one last time on the rocks of the threshold.

Those rocks. Not just in the wall but all over the uphill site. There's the big gray rock with the ants under it. The little round white rock. The dark rock that the oak leaf is on. The rock with chunks of bronze-colored minerals that the beech leaf is on. The lichen-splotched rock that the maple leaf is on. Put a shovel in the ground here, and you're bound to strike the rounded edge of a rock. As a settlement-era crop farmer, you'll need to haul these rocks off to the edge of your field, piling both big and small into a stone wall. As seasons pass and the ground goes through cycles of freezing and thawing, expanding ice below will drive new rocks up through the dirt onto the surface of your field. Removing the new frost-heaved rocks grown amid your crop will be an ongoing struggle. Maybe it's better just to graze cows and sheep here. You only need to use the choicest big rocks to build the wall that pens your animals, perhaps topped with chestnut posts supporting barbed wire.

Whether cropland or grazing pastures, historic farmland across the Northeast is delineated by stone walls (fig. 4.4). Every wall holds a mixture of types of rocks—combinations of schist, gneiss, granite, marble, conglomerate, and others. These aren't local rocks—they didn't come from the bedrock below. Instead, these stones are fragments of mountains and hills north of here. They were scooped up by the glaciers, carried along for centuries as the giant ice tumbler smoothed their sharp edges, and then abandoned when, in the warming climate, the ice turned to water and crept away. When the mile-thick sheet of ice let go of the rocks suspended in it, they fell to the ground. The rocks in these stone walls are from the layer of glacial till—the footprint of the glaciers—that now coats the whole glaciated landscape of the northern United States. In this area our till is typically about five feet thick, but it varies locally, with extremes of 200 feet thick.

Figure 4.4. A typical
New England stone wall.

We expect to see glacial till everywhere in the Northeast, so
exceptions are interesting. That's why Bark Hollow, with its bedrock peering through, was exciting. That's why the lake delta at
Maggie's Forest, devoid of rocks, was exciting. And that's why this
swamp is exciting. Nowhere do we see rocks. No stone walls, no
stones on the ground, no stones below. Just clay, as deep as the
students can dig. Why is there clay here? The same reason there's
sand at Maggie's Forest—this is the other half of that story.

Where are we? The map puts the Great Swamp elevation at 200
feet above sea level—100 feet lower than Maggie's Forest. Our elevation is near the bottom of the rift valley—at the bottom of the glacial lake. The bowling balls fell out up in the delta, but the sheets
of paper drifted on out to the middle of the lake. Here, they settled
to the bottom, forming layer after layer of clay.

And the uphill site? I like to imagine it as a little island in the
glacial lake, poking up above the waterline. Alas, the top of the hill
is still below the top of the lake delta, and thus below the lake sur-

face. It would have been more like an underwater hill onto which clay did not settle. At least my intuition says that clay wouldn't settle on such an underwater hill, and the geologists I spoke with this year all seemed to buy my theory for this site. But these were vague sentiments, not the sort of authoritative basis on which to found a chapter in my book. So it was time for an experiment.

Down in our basement this winter, the kids and I suspended clay in water. As we mixed, water particles interjected themselves in between the sheets of clay particles. The slurry grew thicker, and after about ten minutes dark slime oozed between our fingers as we closed our fists. Then, when the clay was fully suspended in water, we poured it into a large clear tub, half-full of water. In this large tub we had prepared a whole underwater landscape of various-shaped rocks, rock towers, and rock shelves. Nearby, a small fan simulated gentle winds on the surface of our miniature glacial lake.

As we poured in the clay mixture, our whole lake grew cloudy. But the cloudiness began to settle out as a dense layer of fog near the bottom. A heavy fog that sank down slopes, spilled over edges, and snuck around corners seeking out the lowest point. Before it settled out on the bottom, the bulk of the clay flowed downhill in this form.

After weeks in our basement, when the last of the clays had finally left the water column, the deposit of clay formed a horizontal surface at the bottom of our lake. Any underwater rock higher than this clay floor stood out as an island, with only a minimal dusting of clay on its surface. We experimented by alternating daily additions of red and gray ceramic clay to see layered deposits, and with clay formed from dry cat litter in another lake. When we finally drained the lakes, I left our basement convinced.

The hill in this chapter is the little underwater island I drew in the last chapter, figure 3.10a. Perhaps when the lake drained, for a short moment this hill lived as a true dry island. Today, with no clay mask, the hill retains the characteristic trait of almost everywhere else in the North: a surface covered with glacial till.

In the swamp below, clay hides the till. This is the clay that the *Field Naturalist* students were eagerly digging through. Under the pin oaks, swamp white oaks, and tupelos, a giant dish, crafted by a glacial potter, holds a swamp-sized serving of water. It is filled and refilled by rain above. The tightly packed particles in the dish's clay prevents the water from leaking out. Without the clays the water might just percolate down into the earth, leaving the surface dry like at Maggie's Forest. Wetlands form in different ways, often connected to the water table belowground. But this wetland is "perched" above the water table—it doesn't depend on flow from nearby groundwater, it just sits like a birdbath above the fray.

For 10,000 years this swamp held water. Then in the late 1700s people went to work trying to drain the swamp. Back before we knew that draining a swamp was a bad idea, they cut channels—like the straight-walled creek we crossed on our way in—to let the water flow out. Before the public understood that wetlands prevent flood damage and filter pollutants, the governor established a board of commissioners to construct a "great drain," levying taxes to support this work. Before our society valued the myriad species that specialize in wetland ecosystems, landowners went to work cutting down the trees, hoping to convert the swamp to viable farmland. In this way half of the world's wetlands have been lost. But here, this clay was too deep and the swamp too determined. Here, the swamp persisted.

MAJOR LESSONS FOR INTERPRETING A LANDSCAPE

- What is the spatial context of your site? Think both at a large scale (e.g., in relation to other sites) and at a smaller scale (e.g., from the perspective of a mammal looking to mark territory).
- Consider the origin of rocks and soils at the site.
- Consider subsurface hydrology.

West ← **460 ft**   **310 ft**   **280 ft**

North

GPSmap 76CS x

GARMIN

235 ft

0 ft ➡ East

5. Change

We drop the canoe in the water. Juno steps in, followed by naturalist Julia Blyth, and then me. We float. Under the water, long arms of pondweed, waving in the current, brush the bottom of the boat as we drift downstream. A flotilla of ducks lean their heads forward as they strain to paddle away from us. A dead fish rocks belly-up. A great blue heron glides overhead and drops down toward the shore. The heron extends its legs to the ground and pulls its head up as its huge wings ease the landing with a few final flaps. As we go south, the sun is climbing on our left. In my head is a quote, of unknown origin, which my friend Megan always had pinned to her email signature: "If you want to ease your mind, take it down by the river."

Both Maggie's Forest and the Great Swamp were created by a glacial lake. We are now paddling on that lake—or what's left of it. When the dam burst, the lake shrank and shrank until it was just a sinuous line snaking down the old lake bottom. This river, the Connecticut River, is the remnant of the lake, and we can trace the thread of water, continuously flowing, all the way back 15,000 years (fig. 5.2).

On our right we pass a steep bank full of holes. The bigger holes, the size of softballs, must be the entrances to kingfisher nests. The smaller holes must be made by bank swallows. We hug

Figure 5.2. *Field Naturalist* students paddling on the river.

the shoreline as we drift downstream looking for animal tracks in the wet soil. At the base of the cliff, drag marks in the sand tell of beavers leaving the river, searching for food, and carrying it back into the water to eat. At the top of the cliff, the ends of corn rows peer over the edge, tempting hungry beavers (fig. 5.3). A few weeks ago I fought my way through that corn seeking today's destination, only to find myself lost in a huge, flat cornfield, exposed to a lightning storm, a half mile from the river, and a half mile from my car. I decided a canoe trip would be easier.

As we float downstream, the corn on top of the cliff gives way to big trees. Deadwood strewn about the base of the cliff forces us to steer the boat further from shore. Long, straight trunks as wide as our canoe angle gently down, their tops disappearing under the dark water. Sun-bleached limbs bigger than me pierce up through the surface. Some trunks bend out of the water only to reenter six feet on, like the body of a mythical sea serpent. The most recent victim—a great cottonwood lying by the shore with its head sub-

Figure 5.3. A view of the riverbank showing the cornfield, signs of beavers feeding, and bank nests of kingfishers.

merged—still clings to its furrowed bark. At its base, the trunk expands into an eight-foot-wide root ball, washed clean of dirt, like the frayed end of a giant rope.

Amid these fallen trees, a ten-foot-long mound of smaller sticks marks a beaver's home. A lodge. Beavers are famous for their dams, but on big rivers and lakes, there is no need to build a dam. Dams are for small streams where the water isn't deep enough. Here, the water is plenty deep. They need only build the lodge. In this case, it's a bank lodge—a hole dug into the dirt near the shore and covered with sticks. I can't look at a beaver lodge without seeing that one near the barfed-up voles.

A BEAVER LODGE

Fifteen years ago, in a winter animal tracking course, the instructor led Charley and me through oak forests down to an old beaver meadow. Once, the whole meadow was standing water, an area flooded when the beaver placed a dam across the stream below. But sometime in the year before we arrived, people took chainsaws to the dam and breached it, draining the pond. When we got there it was a big open field in which coyotes would hunt for small mammals. That day, the tracks in the snow showed that the coyotes were catching voles, much as I've seen coyotes doing on the grassy edges of roads and highways. We followed the coyotes to a major intersection marked with many variously aged scats. Assuming that the coyotes would mark the area appropriately, we looked for fresh scat. What we found were the barfed-up remains of many voles. The volume of voles spoke to how quickly the little rodents multiplied and took over the meadow that, not too many months prior, had been a pond. As to why they were barfed up, we still have no good explanation. Barfed-up shrews might make some sense. Shrews seem to taste horrible to mammals—perhaps it's the venom in their saliva, or just their stinky body odor—and so predators often grab them hoping for lunch but promptly spit them out before swallowing them.

Pondering the voles, we wandered into the center of the beaver meadow and found the beaver lodge. Unlike a bank lodge, beaver lodges that are constructed in the middle of a pond are a work of pure sticks—no shoreline is involved. But like a bank lodge, the entrance is underwater. That's why beavers like deep water—so that the door to their home is hidden from marauders. The lodge is tall enough so that, although the entrance is underwater, the rooms inside are high and dry. Now with that pond drained, the entire lodge was dry. The foot-high door stood out in the open, beckoning all to enter. Among the fifteen students and one teacher, only I was skinny enough and foolish enough to succumb. I dropped to my knees and cautiously poked my head in.

Built by first piling up a heap of sticks and then chewing a hole through it, the outside of a beaver lodge is a disorganized jumble of lines, like a two-year-old's scribbles on paper. The pointy sticks, many sharpened by the two-inch-long incisors of the giant aquatic rodent, angle menacingly outward in all directions. As I entered the lodge, I braced myself to be speared a thousand times over. I wormed my way up through the hallway, just big enough to squeeze through if I stretched my arms in front of me. In the darkness, my hands found an opening to the right near the end of the hall. I turned, following my hands further upward into a wider space. In the round room, I could now spin in tight circles, sit, and admire the walls.

Nothing poked me. Not a single stick protruded to snag my clothes. All the walls were perfectly smooth. Cooped up in that room for hours each day, the beaver family sat, discussing the thickening ice on the pond, news of the otter, whether their diminishing store of branches they planted in the muck at the bottom of the pond would last the winter, the ring of distasteful white pines that increasingly encircled their pond as they cut down the choicest hardwood trees, their hopes for expanding the system of terraced ponds further upstream with new dams next year, their dreams of the pond far downstream where the elder beavers had been born, and the legends of the Great River even farther downstream. All the while, the beavers idly chewed any errant bumps on the walls, wearing them meticulously smooth. I have toured great temples, churches, castles, caves, cliff dwellings, pioneer homesteads, and transcendentalist cabins. But none brought me the exhilaration of entering the house of a beaver.

THE BEACH

The steep cliff alongside the canoe fades, gently lowering and receding. The high vertical wall transitions to a low ramp, then into a long, flat expanse almost level with the water surface. The

Figure 5.4. A broad view of the river with the black triangle indicating the location of this chapter's site. (MassGIS, Sanborn LLC)

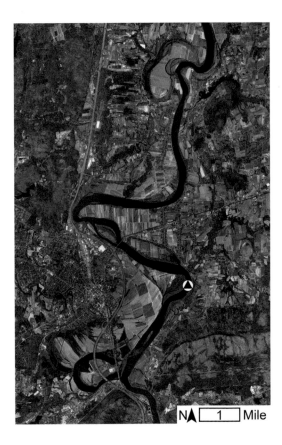

N▲ ☐ 1 ☐ Mile

trees that once loomed over the water's edge are now 200 feet away, on the other side of an open sandy beach. We have arrived. This is Rainbow Beach (fig. 5.4).

We steer toward the beach, lurch to a stop, and step into the shallows to drag the canoe, sand grinding against hard plastic, clear of the water. Juno follows and we prepare to eat lunch and explore.

With the topographic map layer loaded into the GPS, I set it down in the sand by the water's edge and take a picture of the screen where a little black triangle shows our location (see fig. 5.1). Happy with our place on the map, I stand and look around. To the east, on the far side of the river, is a steep, muddy bank. To the west, on the far side of the beach, is the edge of the forest. I drop one end of a measuring tape and walk west. Doing my best to maintain a constant bearing, walking past a few scattered cottonwood saplings in the sand, I measure the distance from the water to the forest edge: 210 feet. At the edge, scattered sandbar willows transition into taller grasses and cocklebur and then into overhanging saplings of black willow and silver maple. Julia and Juno join me in the forest, and we decide that Juno will be a good scale bar for the pictures.

At 235 feet from the river, Juno stands in a dense thicket of short silver maple saplings, most no thicker than his wrist. At 280 feet, Juno stands between a young boxelder and a young silver maple, each about as wide as his two legs together. We walk past some small elms, and then, at 310 feet, Juno feels the furrowed bark of a respectable cottonwood, definitely wider than his body. To avoid stings, we have to carry Juno over the tall wood nettles to 460 feet, where, after I remind him of the protein-rich nettle tea we once made, Juno relaxes against a massive cottonwood, wider than the three of us combined. He looks out at some massive silver maples beyond while I crouch under a boxelder. Why do the trees get bigger the farther you go from the water?

While we are measuring, my batteries run low and I dash back toward the canoe for more. I reach the forest edge and begin racing across the sand. But then I stop. A little dark blur flits by and rests on the sand near me. My heart races. This is atop my list of species to photograph. I approach, and it flits away again to another spot. I fall to the ground and slowly creep up on it, trying to keep my shadow small and my camera out of the sand. Eventually, I get close enough to get a couple pictures. Grinning confidently, I stand up, the creature flees, and after grabbing fresh batteries, I hurry back into the woods.

It isn't until a week later that, seeing little dark gaps in the meandering white paths along the edges of the tiger beetle's wing covers, I realize the creature I was chasing wasn't the right species. According to the field guides, in my picture sits a bronzed tiger beetle.

It isn't until mid-winter that I finally see the federally endangered species I'm after. But she's dead—she died one day ago, housed in a little plastic shoebox in a government facility. Yesterday, as Rodger Gwiazdowski put it, she was "the only adult Puritan tiger beetle on Earth." Now there are none—which is as it should be. Puritan tiger beetles don't overwinter as adults—only as larvae. As adults, the lucky ones only live two months. But in Rodger's cozy laboratory, adult beetles can hang on till the ripe old age of five months.

When Juno, Alder, Sydne, and I shuffle into Rodger's laboratory mid-morning on a Sunday, it's 18°F outside. We're across the street from a big quarry that digs sand from one of the deltas on the old glacial lake. Opening a metal door, we walk into a cavernous warehouse—warm, dingy, and mostly deserted. Around the room, gray waist-high concrete walls delineate large oval holding tanks for aquatic research. One of these tanks has been converted into a brightly lit laboratory, surrounded by white walls and big glass windows. Inside this shining bubble sits Rodger and his little beetles. As we arrive, I joke to Rodger that this building must be where they put the species they've given up on.

The last time I was in this building dozens of Atlantic salmon swam through these tanks. For forty-five years the US Fish and Wildlife Service poured millions of dollars into trying to restore the Connecticut River's salmon populations. Tens of thousands of salmon once journeyed the length of the Connecticut River every year—swimming from small headwater streams and beaver ponds where they grew up, down through the mouth of the river, out past

Long Island up into the Labrador Sea of the North Atlantic, down along the Newfoundland coast, and all the way back again to breed and die in the headwaters.

But then people dammed up the rivers, polluted the water, and started warming the climate. Salmon disappeared from the rivers, along with shad and many of our other fish. In the late twentieth century, we began cleaning things up. We stopped polluting the water. We built special devices—stairways and elevators—to help migratory fish get up over the dams. We started breeding salmon in laboratories and releasing hundreds of thousands of them each year, hoping to restore the populations.

Last year, I took Juno and Alder to visit the fish elevator on one of the major dams in the river. A dam past which the millions of salmon reared in the laboratory and released into the river would have swum on their way to the sea. A dam past which they would have to swim on their return. We entered the concrete-walled hydropower plant next to the towering dam and wound our way through giant gears and turbines up an open metal grate staircase. The space was filled with humming and roaring. Water rushing below the building spun the gears. At the top of the staircase, light poured in through open doors that led out to a viewing deck. The deck was wet from the constant spray of water spilling over the dam and crashing below. Looking down, we saw a few shad desperately trying to swim up the face of the dam and inevitably failing despite an impressive effort.

Suddenly, the enormous mechanical elevator began to move. Chains thicker than my arms hoisted thousands of gallons of water up to where we stood, thirty feet above the river. The giant cube of water was almost within reach when the back of it opened and dumped the payload, wriggling with fish, into a hidden chute in back. We hurried around to the other side to see the fish. In a dark hallway, big windows showed silvery shad and snakelike lampreys, longer than my children, swimming in murky water a few inches from our noses. The fish were soon shuttled through to another room where researchers sat in a private concrete tube, counting.

Later, I looked online for the total counts for the whole season. American shad: 385,930—not bad, considering the number was under 5,000 in 1955, but not the millions it once was. Sea lamprey: 35,249—definitely better than the two counted in 1957. Blueback herring: 137—a catastrophic decline from the 632,255 they counted in 1985. And the total number of salmon that swam up past the dam last year? Exactly three.

Despite nearly fifty years and over $25 million invested in the effort, the recovery of salmon in the Connecticut River failed. The elevators and ladders on the dams aren't sufficient. They don't let enough fish through, plus there are too many small dams up and down the tributaries, all of which need to be removed in order to fully restore the system. But even if we fixed the river, the salmon are still in trouble because we haven't fixed climate change. The Connecticut River salmon swim at the southern edge of the species' natural range. In a warming climate where we expect species ranges to shift northward, you don't want to be among the individuals occupying the southern part of your species' range.

But it's not simply that the river water is getting too warm for salmon to survive. Much of the problem is out at sea. The complicated and poorly understood dynamics of the salmon's marine habitat seem to have been intractably altered. All salmon from New England up through northern Canada come together to feed and overwinter in the North Atlantic. There, circulating water is driven by the interaction of cold, fresh water carried in by the Labrador Current, which mixes with warm saline water brought up by the Gulf Stream. It's a complex system, and small changes can have a big impact. In a warming sea, it seems that changes in the plankton on which the salmon feed, and possibly in the abundance of predators, has made these places—the Labrador Sea and the Grand

Banks—less hospitable for the salmon to feed and find winter refuge. Thus, no matter how far we restore the rivers, Atlantic salmon populations are declining everywhere.

Although it failed at its goal, the salmon recovery effort wasn't a complete waste. Restoring the rivers for the sake of salmon helped a lot of other species that also depend on clean, connected water. Today, when the small shadbush trees bloom, we can once again find hundreds of thousands of shad migrating upstream.

TIGER BEETLES

In a giant concrete tank where salmon once swam laps, Rodger now sits in front of stacked cases of miniature beach habitats built for tiger beetles. I'm reminded of the catastrophe that enveloped another of Rodger's experiments back in graduate school. One afternoon, he meticulously checked that each of the dozens of federally threatened northeastern beach tiger beetles the team had been raising all summer were fed, safely burrowed into their holes in the sand, and closed into their respective containers. He then sealed them into the environmental control chamber—a refrigerator-sized unit that precisely maintains perfect temperature and humidity—and went home. But overnight the brand-new unit went haywire, raising the temperature to over 120°F. The next morning Rodger found shelves of toasted larvae, months in the making, the entire captive population of this protected species. That was years ago. For the current project, Rodger brought on an engineer to build their own environmental control chamber, with remote monitoring.

In his new lab, Rodger is perched on a stool, lanky limbs on a wiry frame, sporting a green down vest over a patterned sweater and round-lensed glasses with thick black rims. He's surrounded by images of Puritan tiger beetles, whose metallic green wing covers are patterned with white around the edges, whose black eyes bulge from their heads, and whose bodies hover atop long legs meant for running across hot sand. Clearly some comingling of styles.

As we catch up on times since graduate school, I fill Rodger in on my book project and ask if he knows why the trees at Rainbow Beach get bigger the further you go from the shore. But Rodger's beetles don't venture into the forest, so neither does he. Still, I think the answer to the tree mystery must be connected to the beetle management.

I lift a large plastic tub off a cart in Rodger's lab and put it on the floor for Juno to inspect. Behind Juno is another cart loaded with crickets and darkling beetles destined to be meals for the tigers. In front of us in the tub are dozens of clear plastic tubes, each the size of a roll of quarters, standing up on their ends. They are half full of fine, slightly damp sand. On the surface of the sand in each tube sits a dark hole just big enough for a grain of rice to slip down. Some tubes have two holes, some have none. Picking up a tube and looking through the clear side, we can see where a hole becomes a tunnel as it worms down the container edge into the sand. Somewhere in that tunnel lurks a young larva. In another couple years, it will be big enough for Rodger to release into the sands of Rainbow Beach.

In nature all you see is the top of the hole—dark, round, apparently featureless—perhaps just the mark where a child stuck a small pencil in the ground and pulled it out. The lack of features around the hole is itself the distinctive feature of a larval tiger beetle burrow. Unlike an ant hole, piled with soil excavated from below and dumped at the burrow entrance, the rim of a larval tiger beetle hole is clean. You'll find no debris on the ground surface for an inch or so in all directions around the hole. Look just beyond this cleared zone, and you may see a miniature hill of soil thrown there by the larva (fig. 5.5). The larva scoots to the top of the burrow, carrying a tiny clump of soil. With a quick flip of her head,

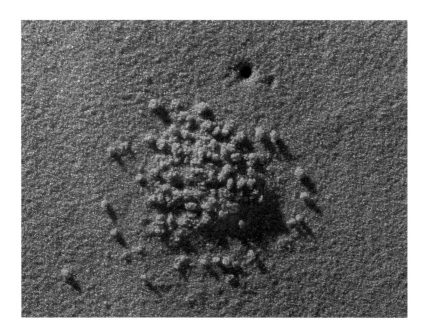

Figure 5.5. Tiger beetle burrow in Coral Pink Sand Dunes State Park, Utah, showing the characteristic beveled edge of the hole and the throw pile.

the soil is sent in an arc over the clear zone and lands neatly in the growing pile an inch or so away. If you are feeling mischievous when you approach an open hole, drop a couple grains of sand down it and see if she flings it back out at you.

Sometimes, if you look closely, you'll notice that the very edges of the open burrow entrance are gently beveled. This is where the larva will rest her flattened head, the same color as the surrounding soil, fitting tightly as it closes the burrow like the lid of a pot. Or perhaps more like a cork in a freshly shaken champagne bottle, with her little eyes peering upward anticipating lunch. From the vantage point of the poor little ant or baby cricket out on a stroll, nothing betrays the trap. The path ahead is free of obstacles, meticulously so. In a flash the ground transforms into a raging tiger, and a tiny life is taken.

Puritan tiger beetles, like the Atlantic salmon, are in trouble. At least with salmon, even if they go extinct in the United States, there are rivers in countries all around the Atlantic Ocean that still support members of the species—in Canada, Greenland, and most of Europe. There are even places outside of their native range where Atlantic salmon have been introduced—the Pacific Coast of the United States, South America, New Zealand, and Australia. But Puritan tiger beetles are found in two small areas: in part of the Chesapeake Bay in Maryland and at a couple sites along the Connecticut River. And Rodger believes that, based on differences in behavior and genetics, these will soon be recognized as two separate species. So really the species I'm looking for on Rainbow Beach occurs primarily on just this and one other little beach along this one river.

Not too long ago Puritan tiger beetles could be found on at least eleven beaches up and down the Connecticut River. And back then it would have been much easier to find them here at Rainbow Beach. What went wrong? I put the question to Tim Simmons, a restoration ecologist who has spent decades managing these critters' habitat for the state wildlife agency. It's mid-winter when I pick up the phone to call Tim. Through the crackle and hiss of the speakers and the interference in the air, Tim explains his long history with the beetles while I sit in my office, staring though my window at the snowpack on the ground.

Outside on the river the frozen surface—a polished skating rink in places, a pile of shattered pieces in others—pops and hisses as it moves impossibly slowly, tugging forcefully at the bank. In spring, the ice disintegrates. Propelled by floodwater, a truck-sized block crashes into shore, scours the vegetation and ground beneath, and moves on. The sun rises, the remaining ice melts, and summer arrives. A fresh scar in the land, devoid of vegetation, records the river's fury last winter. In this sunny little spot where the ice tore up the land, the soil is now warm, dry, and finely textured. A tiger beetle, plump with eggs, runs over to this perfect soil, angles herself upright, and, dancing like a miniature jackhammer, deposits

Figure 5.6. Beach substrates in the tiger beetle burrowing zone on Rainbow Beach. (Juno Charney)

an egg. Over the next two years, a larva fastidiously manages her burrow as she grows. At last she transforms into an adult and explodes out into the sunshine, glimmering with hope.

But life aboveground isn't so easy. As she searches for food and spouses, she is continually interrupted by human children throwing beach balls and oversized men spilling out of their bathing suits looking for somewhere other than their motorboats to relieve themselves. Digging sideways into the warm sand, she constructs a shallow little cave in which she rests and reminisces over her underground childhood. One day, a strange dry fog rolls down from an airplane flying over a nearby cornfield. For the following week she has chills and a sickly feeling in the pit of her stomach, but this passes. Many of her would-be spouses are less lucky.

When it comes time to lay her eggs, she can't find any fresh scars from past winters' ice scour. All around, she is annoyed to find the signs of the unsophisticated bronzed tiger beetles, who seem to be haphazardly procreating in subpar soils, oblivious to the refined and highly selective standards of a Puritan tiger beetle. Eventually, she finds some decent exposed earth in the trampled edges of what Tim Simmons calls the "latrine trail" formed by human beachgoers. As she plunges her butt into the soil, she hopes her little ones find an easier life aboveground than she did. Later that summer, after being buried in six inches of silt left by seventeen feet of floodwater, only a handful of her many larvae are able to dig themselves out and set up ambush for their prey.

What went wrong? A lot of things. For one, a nuclear plant upstream, a series of dams, and a warming climate have raised the temperature of the river such that less ice forms. Though not the only force that creates Puritan tiger beetle habitat, Tim thinks ice was historically important for creating their habitat here on Rainbow Beach (figs. 5.6 and 5.7).

Climate change and the upstream dams have also conspired to create bigger and more frequent flooding of the beach. It's not the existence of the dams per se that's causing the flooding but changes in the timing of when the hydroelectric companies choose to release their water stores. Tim is hopeful that when the power plants renew their operating licenses there will be an opportunity for conservation concerns to help guide future water release schedules. The nuclear plant, too, has been shut down, alleviating another stress on the river. The climate, on the other hand, seems hell-bent on delivering increasingly severe deluges.

Boaters partying on the beach do create some habitat for larvae by disturbing the soil, as sort of a stand-in for ice scour. However, because people on the beach are so disruptive to the adult beetles, Tim believes that beachgoers on the whole are a net drain on the species. That's not even counting the people who are intentionally attacking the beetles. Twenty-five years ago two dentists and a lawyer, who on the side were hobby tiger beetle collectors, went to Rainbow Beach and collected every single beetle they could find. Perhaps we should just close the beach to people. If the

Figure 5.7. (a) Ice scour along a small river, (b) ice jam on the Connecticut River, and (c) ice chunks close to the shore on the Connecticut River.

beetles had vertebrae, Tim thinks that would happen. But bugs, even federally endangered ones, don't hold enough political clout to close even a section of a popular public beach.

Decades ago Tim began fighting against the tiger beetles' enemies. He and his team tried to artificially create habitat for Puritan tiger beetles, swarming the beach with rakes, tilling machines, and bare hands. Mimicking the work of giant blocks of ice, they scoured the land, ripping up vegetation and soil to create perfect sites for larval tunnels. But the successes were always short-lived. The beetles like to live right at the back of the beach, near the forest edge. But every time Tim would create a perfect nesting spot, creeping vegetation or floods would ruin his team's work within a few years. Beach-going people kept coming and harassing the adults. Sometimes Tim would be dowsed with pesticides from the nearby fields. However he tried to help, the fight for tiger beetles on Rainbow Beach was always an uphill battle.

At the core of the problem, rivers are naturally dynamic. Tim realized that his team was, as he put it, "trying to manage static conditions on a place that wants to be violently dynamic. That's a lesson that habitat managers have to learn—usually the hard way." Rainbow Beach has never been and never will be static. The sands and vegetation are constantly changing. Where the tiger beetles want to nest this year won't be the same as next year. The problem is that so much of the river has been impacted by humans, there aren't a lot of other places left for them to go.

CATCHING TIGERS

As the field season approaches, Rodger's team gears up to catch tiger beetles. The plan is to have the adults lay eggs in the safety of the lab, free from predators and competitors, where Rodger can fatten up larvae and ultimately release the offspring back into the wild. I stop in one afternoon to hear a pep talk by Hal Weeks, who is sharing the success story of the Oregon chub—a little fish that depends on the backwaters in wildly meandering rivers in the Willamette River Valley. Humans, with their habit of overtidying things, found these rivers and simplified the system by straightening channels and controlling floods. This effectively eliminated chub habitat, pushing the fish onto the federal endangered species list. Hal and others went about restocking the fish into remaining habitat, working with farmers to protect populations, and convincing the department of transportation to stop spraying near critical waters. After twenty years of work, the fish was deemed "recovered" and taken off of the endangered species list.

A few days after Hal's presentation, carrying a crew of Puritan tiger beetle volunteers, a federal motorboat drops us off on Rainbow Beach. Standing at the water's edge, I distribute maps, a GPS, a compass, and printouts of this chapter's image to the students and volunteers. Turning the map upside down and then right side up again, they're having trouble figuring out why the GPS seems to be putting us on the wrong shore. We eventually solve the map puzzle and then head off to practice catching tiger beetles. Today we're only catching the common species, so that when the rare beetles finally emerge in a few weeks, we'll have made all our mistakes with something not so precious.

As we wander near the back of the beach, I duck quickly into the trees to search for Juno's hat, which we left here last year. It was next to a big wooden display kiosk erected years ago to tell visitors about the endangered beetles. As I enter the woods, I'm disoriented. The saplings Juno had posed next to on our previous visit are flattened to the ground—presumably by the ice jams and floods this winter. The odds of finding his hat look grim.

I muse over my own cherished hat—from the Telluride Bluegrass Festival—that I left on an island off the coast of Florida. Like debris in the water, a flood of memories washes past me. That hat came to me during a mid-college road trip, Operation Monkey Storm, not long after my friends and I were evicted from the Alamo for playing croquet in the courtyard. I insisted that we visit Colorado's Great Sand Dunes, a place that I had fallen in love with at seven on a family road trip, mirroring a trip my dad had taken as a child. My dad had allowed his poky stubble to grow into a proto-beard while my brother and I lost a little metal Band-Aid tin full of model airplanes in the shifting sands of the Colorado Dunes—then returned and miraculously found the tin. After the dunes, Operation Monkey Storm headed to Telluride, where I bought the hat.

For years after Operation Monkey Storm, I dipped that hat into every river I encountered. In a little ceremony, I'd scoop water to all the cardinal directions, then end with one big scoop over my head, letting the water run down my back. The Russian River, the Missouri, the Mississippi, the Connecticut, the Columbia, the Rio Grande, the Saint John, and countless little creeks in between. The hat took on a pungent smell, like mildew and snakes, but always stayed with me. When my dad was dying, I placed the hat on the floor of the room, scoop-side up. Somehow I felt that all the sacred river energy would either flow out into the room and help, or that the hat would soak up whatever sacred energy was floating in the room to save for later. I can't be sure, but I think it helped in some way.

Several years later, I was with Liz Willey and Mike Jones—my turtle friends—out on an island off the coast of Florida searching for turtles with my threadbare hat. The island had been made by Calusa Indians thousands of years before, entirely out of discarded sea shells. At the time that it was made, the mound of shells may well have been connected to the mainland before sea levels rose and turned it into an isolated island.

On that island, I found myself in a thicket of barbed wire cactus—a maze of long, spindly arms that swung through the air with two-inch spines. The spines kept going right through my layers of clothes and flesh until they landed with a thud against my bones. Several tips broke off in my fingers. Two years later, hiking on Stewart's Island off the southern tip of New Zealand, one of those spines reared its bloody head from my thumb, taunting me. I had no tweezers, and it slipped back in, where I can still feel the lump to this day.

I found a turtle in that Florida thicket, but when I emerged, I realized my hat was gone. Despite searching, I couldn't find it. A year later Mike and Liz returned to the island without me, hoping to encounter the hat while they searched for turtles. But it remained lost.

In the woods behind Rainbow Beach, I find the wooden kiosk. It's covered with flyers about the tiger beetles and carefully worded messages imploring visitors to respect their habitat. But it's also completely overgrown by trees. Hidden back here, there's no way any beachgoers would ever see the board or get the message.

Two years after I lost my hat in Florida, Mike and Liz returned with our friend Derek. About to board the motorboat for the return home, Derek decided to make one more quick run out after a turtle. A few minutes later he returned to the boat. In his hand was my hat. Or what was left of it after rotting in the Florida Everglades for two years—primarily the brim, the spider-like network of seams, and a few pins and porcupine quills I'd stuck in it over the years.

Clinging to these memories as they slip past feels like scooping water with only the skeleton of my hat. How does the philosophy go? I can never dip my toe in the same river twice because it's a different river from moment to moment? I never die because I am a different "I" from moment to moment—or rather, I am constantly dying and being reborn in each moment? The river and I don't exist as discrete things but are merely two overlapping heaps of moving parts?

Staring into the forest behind Rainbow Beach, I suppose Juno hasn't lashed his soul as tightly to his hat as I did to mine. It's probably been washed far downstream. Maybe we should just go back to REI and buy a new one. Turning my back on the forest, I return to practice catching the common tiger beetles.

On a sunny July day a couple weeks later, I return with a different federal team. This time, the beach is teeming with real, live endangered Puritan tiger beetles. I belly-crawl through the sand trying to capture portraits before they flit away. Hanging out on damp sand near the water, the beetles in my lens crawl with their bellies also near the ground to stay cool. Then I find one in the hot, dry sand. Perched atop a four-inch dune, the beetle fully straightens its six long legs, stilting its body as far as it can away from the radiating heat below. Classic tiger beetle behavior. This is how they have adapted to hot, sandy microclimates. My camera clicks once, then the beetle flits away.

A bit further up the beach, we find a Puritan tiger beetle approaching a dead wood turtle, not far from sandbar willow—all three species listed under the state's endangered species act. I text a picture of the turtle to Mike, who has been following hundreds of individual wood turtles for decades, keeping careful track of each one's idiosyncratic life. Often, when you find a wood turtle around here, it's one that Mike knows well. This one, it turns out, isn't one of his.

FOREST EDGE

I wonder, is saving one species at a time really going to get us where we need to be? Rather than going about conservation one single beetle at a time, protecting whole systems often seems like the most sensible approach. After all, there are over 100,000 species of insects in North America, an unmanageable number. The number of discrete habitat types is much smaller. But sometimes single-species approaches are all we can do. For one thing, some of our most powerful conservation laws are the various federal and state endangered species acts. These laws don't recognize endangered ecosystems in the same way that they recognize endangered species. Ideally, we can focus on umbrella species—that is, species whose protection necessitates the protection of entire unique systems. The target species becomes a crutch to advance a broader conservation mission. And, ideally, we can use other proactive measures to protect valuable systems.

When Rainbow Beach was originally purchased as a conservation area, it wasn't for the Puritan tiger beetle—nobody even knew the beetle lived here. It was actually for the forest behind the beach. This is a rare example of floodplain forest. Silver maple. Cottonwood. Boxelder. These are wetland trees that particularly love rich riparian soils. But riparian forests are under threat. There aren't many left, and the ones that are tend to suffer.

Assembling images for this chapter, I realize I don't yet have a good picture of the smallest trees in the small-to-large gradient. So I dress my scale bar in his orange-and-white shirt, black shorts, and a fresh haircut, and we head back out to Rainbow Beach.

On this visit we're accompanied by Joe Rogers: geologist, river ecologist, and the town's assistant conservation planner. Joe, who stands at six feet, six inches, folds himself into my car, and we drive along the runway of a small airport, bounce on dirt roads through a vast plain of cornfields, and park at a rusty metal gate shrouded in poison ivy. Joe thought he had the key that would allow us to drive right up to the edge of the beach. He doesn't. Instead, we duck under the gate, drop down a short slope, and land in another flat cornfield.

For the next half mile, we fight our way through eight-foot cornstalks. Joe strides ahead, lost in the sea, and I feel like I'm drowning amid the crashing sounds of corn leaves slashing at my ears. Juno clings to my torso, shielding his face from the leaves' serrated edges, and I worry about what pesticides the corn hurls at us. When I'm almost ready to give up, we arrive at the western edge of the forest. On the far side of this forest waits Rainbow Beach. As we stand with the corn behind us and the forest in front of us, we hesitate. We are staring at a tangled wall of blackberries, bittersweet, and poison ivy supported by a frame of boxelders. If we tear a hole through this wall, will it be easier to walk in the forest interior? There is only one way to find out.

Forest edges are rarely inviting. Often a vertical curtain of vegetation seals off the forest perimeter. Outside the forest, something—lawnmowers, cattle, water, rock outcrops, asphalt—prevents woody shrubs from growing. Inside the forest, shade limits the growth of shrubs. Right at the edge, however, there is plenty of light and nothing to prevent the growth of woody stems. The plants respond by filling in the space with leaves—it's almost like the top of the forest canopy turned on its side.

In the fabric that forms edge curtains, sometimes the species are harmless enough: forest shrubs like spicebush and mountain laurel; native North American vines like trumpet creeper and grape; and the saplings and side-shooting branches of the canopy maples, oaks, and pines above. But many of the species can be downright threatening: multiflora rose, whose thorns, curved like sharpened cat claws, are quick to draw blood and tear clothes; poison ivy, whose oils once turned my legs into two long cuts of seared steak requiring steroids to heal; blackberries, whose prickers are not as menacing as rose thorns but can still tear at my jacket and inflict pain on my skin; and a whole suite of invasive species like bush honeysuckle, bittersweet, wisteria, and tree of heaven that threaten our native ecosystems.

To appreciate the forest, you've got to get past the edge. It's deep inside that the true nature of a forest reveals itself. It's in the forest interior that the shrub density is usually thin enough that it's easy to walk and see the grandeur of the trees. Here in the interior the forest is beyond the reach of the extra sunshine, the extra wind, the noise, and the pollution that pours in from the edges. It's beyond the reach of many of the exotic species, which are often concentrated at the edges because that's where the birds, attracted by the higher density of berries and bugs, pooped out the exotic seeds. The interior is beyond the reach of the predators that live out in the field beyond the forest and will venture only a short distance into the forest edge to feed. Many of our sensitive species—small salamanders, skittish birds, delicate ferns—can only survive deep in the forest interior, hundreds of feet from the edge. That's why, as a conservationist, I fight to protect big patches of forest with low perimeter-to-area ratios.

TERRACES

Carefully slipping around poison ivy, Joe, Juno, and I enter the riparian forest. Immediately, the ground drops us down a short slope to another lower level. This is the third shelf we've encountered on the system of floodplain terraces: we started on the plain of the airport, then we stepped down into the cornfield, and now we are down where the riparian forest grows, about fifteen feet below the airport level. Why, I ask Joe, are there terraces along rivers? Whether it's a big river or a small creek through the woods, the water is often flanked by land formed into the shape of a giant staircase.

I understand that each terrace represents a former floodplain of the river—that's the level that the river used to be at thousands of years ago. And I know that the river has been working hard to cut down through the layers of sediment dropped at the bottom of the glacial lake. This down-cutting is given extra speed because, after the glaciers melted away, the continent lifted upward, free of the heavy burden of ice. The river, responding to the uplifted land, is driven to cut ever faster down into the underlying rocks and soil. So, naturally, the current floodplain level will be lower than previous floodplain levels. But why are there discrete steps? I imagine a gradual process of uplifting land and down-cutting water. Wouldn't that produce a continuous slope from the airport down to the riverside?

"It turns out," Joe suggests, "things don't happen gradually, things happen episodically . . . big change happens during big events." For a long time the river might calmly meander back and forth at one level. But then an enormous flood arrives, quickly reshaping the landscape. Braving the storm, gawkers might drive to the edge of the river and marvel at how the top of the water is now thirty feet higher than normal. But beneath the surface the bottom of the river might also be suddenly much lower than normal. The force of all the rushing water mines away tons of old sediment in a flash, carrying it out to sea. The flood recedes, and now the river suddenly finds itself sitting on a new lower level, having abandoned its old floodplain as a terrace up above.

Or, in a similar vein, perhaps the river is slowly cutting downward, and then in a big storm suddenly shifts sideways to a new position where it resumes downcutting. Such a start-and-stop migration can explain a series of staircases.

These explanations aren't too different from what the geology textbooks say. But most textbooks seem to suggest that the different terrace levels indicate larger-scale periods of stability and change. That is, for a while everything is stable, and the river forms a nice flat floodplain. Then there's a prolonged period of change—glacial rebound, tectonic uplift, falling sea level, or a climatic shift—which causes the river to start cutting down until another period of geologic stability occurs. The textbooks point to paired river terraces, frequently found to symmetrically flank a river to indicate the old floodplains during periods of stability.

I suppose that's plausible, but . . . I don't know. Despite being backed up by over a hundred years of literature on terraces, that explanation doesn't feel terribly satisfying. It still seems like, at the scale we're talking, the downcutting should be a gradual process.

Tekla Harms, my former geology professor, suggested another, satisfyingly simple explanation for terraces—an explanation published back in 1909. I visited her this year looking for answers, and she unrolled a big map of the valley on her desk—made of old USGS topographic sections that had been taped together. The current river ran down the center of the map, surrounded with Tekla's notations. On the inside bend in the river, she pointed out topo lines marking a series of low undulations called "scroll-bars"—remnants of past positions of the beach fronts as the river migrated. Tekla then traced her finger along a broader set of contour lines that she'd long ago highlighted on the paper map. East of the current river position, these tightly clustered lines ran parallel to each other as they snaked up the map. They indicated a long, low ledge of soil—the outermost river terrace. The flat land above, just east of the ledge, was the bottom of the old glacial lake. It remained more or less as it was when the lake drained 12,000 years ago. Sometime in the past the meandering river had made its way just to this ledge, cutting through the old lake bottom as it moved eastward. Then, on a whim, the river decided to wander back to the west side of the valley, leaving the terrace behind.

Time passed, and as the river meandered through the valley, it cut deeper downward. Its floodplain got lower and lower—yes, gradually. Then a curve of the river swung to the east again, back near that first terrace, although, by chance, it didn't reach all the way to the old terrace. Instead it shifted back to the west again. Because so much time had passed between the two times the river found itself on the east side of the valley, the level of its floodplain was substantially lower the second time. Thus, the second terrace it cut was at a lower level. The steps in such terraces simply indicate the time that passes as the river swings back and forth in the valley bottom.

The beauty of this explanation is that we don't need to invoke external agents or particular conditions. Terraces simply arise from the intrinsic nature of meandering rivers. But then would terraces on either side of the river be paired? Well, according to recent computer simulations, yes. Because often the river meanders in such a way as to cut off terraces on both sides of its active channel before it has much time to cut downward.

Beyond not understanding terraces, I also never fully understood why floodplains are so flat. I know that water wears down a landscape, and water sits flat on the land—so the level to which it cuts will be flat. But I failed to understand that water also flattens by filling—in the same way that the clay slurry in our basement experiment settled into the deepest spots of our plastic tub to create a horizontal surface. If you look at the profile of the bedrock underneath the Connecticut River, it's anything but flat. Strip away all the soil, and there's a dramatic landscape of peaks and valleys down there. There are places where bedrock still protrudes in the river and spots where you could dig for over 200 feet and hit nothing but muck. It's just that the river and glacial lake in their turns came along and filled in all the holes. Now all we see on top is the flat surface good for farming and landing planes.

RIPARIAN FOREST

Inside the forest, Joe, Juno, and I find a dark, tangled magic. There are few shrubs—mostly big trees above nonwoody vegetation. But making forward progress toward Rainbow Beach is still not trivial. We crawl under and over huge fallen trees. The land gently undulates up and down, and in the muddy low-lying clearings that recently held water, it is easy to navigate. But in other places the wood nettle is thick and chest high, forcing me to hold Juno up above my shoulders so that he clears the nettles. The nettles bite at my elbows, and when we move too quickly, the thicker needles along their stems pierce through my pants. Soon my skin is marked with itchy white welts. Though Joe's arms tower above the nettles, his pants are much thinner than mine, and together we yelp at the stings.

Still, I love wood nettle. You only encounter it in very specific types of places—moist, dark, rich. It's a good native species that supports local wildlife and reminds me of shady streamside forests of my childhood. Legend has it that a little sting from nettle each year even helps with arthritis. As a kid, I would occasionally run through the center of nettle patches just to experience the thrill of the short-lived stings. Today, however, my increasingly pained skin is happy when we find a seven-foot-tall thicket of Japanese knotweed smothering everything beneath it in darkness. There we crack hollow knotweed stems beneath our boots as we plunge ahead.

Despite walking hundreds of yards, it feels like we never really get far from an edge. Everywhere we go in this forest there are signs of disturbance—recent floods, sometimes carrying damaging blocks of ice, have ripped away at the inhabitants. Into the light-filled gaps created by these disturbances come invasive species—knotweed, bittersweet, garlic mustard, catalpa, and more. Sometimes it's the floodwater that's brought the species in. Some landowner may have thought she was helping the environment upstream by cutting down a knotweed patch, but the cuttings she left on the bank were carried by the water to this spot in the forest, where they sprouted roots and started a new colony.

Few riparian forests remain, and those that do are suffering. They suffer because, like this one, they are small and frequently disturbed. Bounded on one side by cropland, on the other by a river, this is just a thin fragment of the riparian forest that previously covered most of the floodplain terraces. Once riparian forests up and down the river were connected to each other and to vast contiguous forests into the uplands. Sensitive forest dwellers could easily roam long distances without leaving the safety of the trees. Now this small patch isn't connected to anything and is assaulted by edges on all sides.

In truth, it's a miracle that this forest is here at all. I wonder how it slipped through hundreds of years of agriculture. The flatness of the land, combined with rich soil and a lack of big rocks, makes for extremely valuable farmland. It's rich and rock-free, in part because we're sitting on the old clay-settled bottom of the glacial lake. But the soil is also renewed in every flood. The rushing floodwater surges over the riverbank, spilling across the land. Then it stops. While water within the river channel keeps rushing down to the sea, the water up in the floodplain stands still. The stillness of the water after the flood is what allows the nutrient-rich clay particles to fall out and settle across the land, replenishing the soil.

Eventually, Joe, Juno, and I, traipsing up and down over the old forest-covered scroll-bars, reach the south end of Rainbow Beach, where the shore is just a thin strip of mud. We wander northward, and the beach widens with every step. We tread across strange long runners from invasive reeds creeping across the sand. Soon tiger beetles flit away from us, and we stop to admire the drag marks where the beavers have cut willows and brought them back to the river to feed. We chat with a family of boaters about the beaver that chewed off the hoses on their boat one night.

When we reach the spot where Juno and I landed with Julia months before, Joe and I swear never to walk in again: next time, we're taking a boat. After gathering up all of our courage, we begin the long trek back to the car, trying a different route this time. On our way out, Juno makes up a song, "On the Other Side of Rainbow Beach." We pass a troop of college students scattered in the trees amid clipboards and measuring tapes taking notes, learning all that they can from this rare, changing forest.

MEANDERING

Julia, Juno, and I roll up our measuring tapes and head back to the canoe. We walk across the sand—Juno taking pictures of a fish skeleton that he wants to include in the book, and Julia and I discussing tiger beetles. When we get to the shore, we unpack our lunch and maps and talk about the site. Julia picks up the GPS and scowls. The maps and her intuition are offering conflicting signals. The GPS seems to be saying we're on the wrong bank. It's messing with her sense of direction. I know the feeling.

Sense of direction is a funny thing. Sometimes, once it's set, it's hard to undo. One of my first introductions to this valley was when I arrived here on a bus driving west from the Atlantic coast. We exited the interstate and drove east over the bridge that looks out to Rainbow Beach. The bus was driving eastward just before it dropped me off in town. But, having just come from the coast, my body felt that we were still headed west. From then on, my gut instinct in this area was that east was west and south was north. It took me decades of intentionally watching the sun track across the sky to rewire my brain—and still parts of that instinct remain.

Rely too much on technology, and you never develop a sense of direction to begin with. The summer I chased Jefferson salamanders across the Berkshires I had thousands of vernal pools programmed into my GPS and trusted its guidance as I drove back and forth across the county visiting up to ten random sites a day. I've seen almost every part of the Berkshires and have many fond memories of forests and amphibian-filled ponds that I've tromped through. But I have no idea where any of these places are, or how to find them again, other than digging up my old GPS coordinates. When I drive through the Berkshires today, I am frequently surprised as I turn the corner. In front of me is a familiar place associated in my mind with a nearby wetland, but because I never developed a map in my head of where these places are, I never know which site I'll see next.

One of my favorite tools to hand to the *Field Naturalist* students was a little compass I kept on my watchband. My mom gave it to me, but for some reason about half the time it would point exactly 180 degrees in the wrong direction. Some weeks it would be right, but other weeks north was south. The students, of course, didn't know this. With the sun shining down on the landscape in front of them, they had to decide whether to trust their senses or the technology. They already knew not to trust me. I can still feel the wounded glare Grace shot me when she finally figured out the compass trick.

While Juno digs in the sand, Julia and I look across to the muddy cliff on the far side of the river. Clay holds the vertical wall together. It's a classic cut-bank in a meandering river. Up in the mountains, where the rocks are barely weathered, there's not much clay to hold things together. Those fast-moving rivers carry and deposit lots of large rocks and gravel, which don't hold their form. When a river cuts through a gravel bank, it will usually collapse into a low pile. Under such conditions, you tend to get braided rivers made of networks of crisscrossing channels. But on the gentle slopes of the lowlands, where the rivers carry mainly fine particles and there has been plenty of time for weathering to produce clay from granite, you get lazy meandering rivers.

Once the river starts to bend, the bend gets progressively more curved. The water slamming against the outside bank cuts further and further into the clay. Meanwhile, on the inside of the curve, the water is moving at its slowest and drops loads of sand. In this way the whole river creeps into its curve, with an expanding sandbar and a retreating cutbank. Eventually, the bend gets so extreme that the river practically loops back onto itself. And then it actually does loop back onto itself. The river breaks through a final wall and a whole bend is just abandoned. An oxbow, a boomerang-shaped lake slowly filling with sediment, stands as a tribute to an old bend in the river. There are plenty of them on the maps in front of us (see fig. 5.4).

We think of the land as stable, but it's not. Especially not near water. In love with the power and beauty of water, we build along the banks of rivers and oceans expecting nothing to change. When the power of water frustrates us, we build concrete walls and earthen berms to keep the water from moving. We wreck the water's beauty along with the upstream and downstream ecology. Ultimately, the water will keep fighting back. Accepting change is a hard lesson to learn.

With time, this bend, too, will become just another oxbow. But not exactly where we're sitting. The beach must first travel further east. It's that marching of the sand that vexed Tim Simmons—as the tiger beetle habitat steadily crept downriver, off of the land owned by the state and onto the private land of neighboring farmers.

Sitting on the beach, I pull out some historic maps and a piece of clear plastic—a transparency sheet from old overhead projectors. With a 2012 satellite photo behind the plastic, I trace the current banks of the river with a black marker and put a black star where we are sitting. Then I shift the plastic to a map from 1990 and retrace the riverbank with a blue marker. Then I do the same for a 1962 map with a green marker. And finally I trace the river on an 1862 map using a red marker (fig. 5.8).

And there it is. In 1962 the little black star is sitting on the opposite bank. That's about when the USGS made the topo maps that the GPS is using. The GPS has our location on Earth right—it's just that the river has moved more than 500 feet eastward over the past forty-five years.

Figure 5.8. Rainbow Beach in (a) 2018, (b) 1990, (c) 1962, and (d) 1885. Site location is indicated by the black triangle and the 2018 river position is indicated by the green shape superimposed on the images. (a–c: Google Earth; d: USGS)

Roll the clock all the way back to the 1860s, and none of the riparian forest was even on the map—all of that land was created in the last 150 years. That's why the forest is here—the farmers originally did cut down all the forest, but then new land grew in at a lower, swampier, difficult-to-farm terrace, and the forest was left to grow.

Landscapes look so stable and permanent. But even on human timescales things change. The shifting Colorado Sand Dunes swallow our childhood treasures, rising sea water engulfs the shell-mound islands of the Calusa Indians, Earth's crust rebounds from the lost weight of glaciers, meanders cut terraces down into the riparian landscape, the shifting river leaves undulating scroll-bars behind in the forest, ice jams scrape the soil, trees near the eroding edge of the cornfield fall into the water, abandoned hats are washed downriver, and the kiosk at the edge of Rainbow Beach is gradually overgrown by trees.

As the beach we're sitting on moves east, the forest is chasing it. The edge of the forest, with the youngest trees leading the charge, advances onto the sand where the Puritan tiger beetles like to burrow.

And that's why the biggest trees are farthest from the water—because the land itself is oldest there.

MAJOR LESSONS FOR INTERPRETING A LANDSCAPE

- How is the shape of the land changing over time? How fast, and by what means?
- Consider any rare species or targeted management for conservation at your site. Go on to your local regulatory agency's website and look through its maps and guides.
- Where are the habitat edges, and how are they structuring the species assemblages?

Figure 6.1. Large or small, for every pattern there is a process to be discovered.

6. Chemicals

A SHORT TRIP AROUND THE WORLD

I'm not a morning person. Intellectually, I understand that mornings are wonderful—cool air, sparkling dew, singing birds, yadda yadda. But my body sees it differently. I live for night. I guess that's why I study salamanders instead of birds. On road trips, I limp my way through the daylight, straining to stay awake while the sun drags my eyelids down. But once the sun sets, I get a burst of energy and can drive without stopping until the gas runs out. Today I have no choice but to be on the road at 5:00 AM if I am going to catch this field site. Last night I would rather have taken the kids for a long walk through the woods, dimly lit by a crescent moon. Instead we went to bed extra early, leaving the moon high in the sky, so that I could rise now at this decidedly un-amphibian hour, alone.

Low tide is in an hour and a half, and there are 103 miles of pavement between me and the water. Plus I still need to park, unload my gear, drop my boat in the water, and paddle out to the site. Canoe on the roof, I race south. But it's not just the coast that interests me today; this is also a geologic pilgrimage.

Flying past my window are rock faces that were blasted from the hills when the road was cut. Reddish mudstones jut out of the highway median, arranged in flat slabs that are stacked and tilted. In my mind I see the fossil footprint from this rock layer displayed on the floor of the local natural history museum. Next to the display a sign invites Alder and Juno to crawl around in the track itself, pretending to be dinosaurs.

Two hundred million years ago, when this mud was wet on the valley floor, a real dinosaur walked casually along. Her left foot landed here, stayed for less than a second as she moved her right leg forward, then the left foot lifted and, with the sound of squishing mud, she was gone. Impossibly, that instant was permanently recorded. The track captures a tiny flicker of life. It's not so much a description of the animal's build, diet, growth, or ecology—though some of that can be guessed at from the track. What fascinates me is that the track records a personal moment in her story.

At that moment thoughts were in her head. Perhaps her eyes were looking to the left at a friend. Her ears heard splashing in the river nearby. Her belly rumbled from a recent meal. Two rocks clanked together in her gizzard. A pterodactyl shadow approached. In a snap the moment was absorbed into the quadrillions of other moments among the trillions of creatures among the billions of years of our planet. Could she have hoped, or cared, that, after the sun had set and risen seventy billion more times, somehow against all odds that singular moment in her life would be hardened in stone, adorned with decorations, and admired by generations of children?

Dinosaurs aside, after an hour of driving, it's been a fairly tame trip from a geologic perspective. I've been more or less driving south along the center of the rift valley, passing essentially the same sorts of rocks that characterize the northern portion of the valley.

I pass dark, jagged formations of basalt—remnants of the lava flows that filled the spreading rift valley and formed the mountain at Bark Hollow. The road crosses red-brown sandstone—an arkose related to the one at Bark Hollow, but this one, the Portland Arkose, is the famous "brownstone" used to build townhouses in New York City. All these rocks—the arkose, the mudstone, the basalt—are familiar native rocks that were formed here on Laurasia, the supercontinent containing North America, Europe, and Asia 200 million years ago.

Suddenly, the landscape shifts. In just a few minutes, I leave Laurasia behind. The road undulates up and down. Approaching the crest of each hill, I anxiously peer out of the front window to see the upcoming cliff, slowing down as my gaze lingers on the passing rocks. Sometimes I'm able to snap a few blurry, tilted pictures, shooting from my hip through the splattered side window. It seems that each cliff face holds an entirely new set of rocks.

I see dark blocks bisected by three-foot-wide stripes of white. I see gray wavy lines. I see golden flaky layers. I see light gray lenses inside dark gray masses that are streaked with red rust. I see tilting, I see folding, I see intrusions. The patterns stream by in rapid succession, each rock face bringing a completely new story. But all the rocks bear the telltale signs of metamorphism—these rocks were subjected to great heat and pressure, which transformed them physically and chemically.

These metamorphic rocks are the violent scars of continental collisions. These rocks are remnants of the things themselves that glommed on to the front of our continent, like the bugs on my windshield. Several independent landmasses, known as terranes, have been crushed, fused, and swirled together here.

Complex sets of gray schist and gneiss out my side window belong to the Taconic (or "Bronson Hill") island arc. Resembling Japan, this serene volcanic island chain once sat off our coast 450 million years ago. I then drive over rocks that began as sedimentary layers deposited on the floor of the ancient Iapetus Ocean before being thrust onto our continent and metamorphosed into gneiss. Next I'm on to two other ancient landmasses: the Putnam-

Nashoba island arc and the microcontinent of Avalonia. That last landmass, Avalonia, is the most significant.

GONDWANA

Toward the end of our trip around the world, Sydne and I found ourselves as special guests at the grand opening of the Salamander Conservation Center in the mountains of Taiwan. Sitting next to the emissary from Bhutan while aboriginal children danced in red capes in front of us, I tried to fit in by tucking my hair under my hat and patting down my beard—neither of which had been groomed or trimmed in five months. Out of everywhere we'd been in the world, Taiwan was by far the hardest place for us to navigate. Luckily, the salamander biologist Kuang-Yang Lue from the National Taiwan Normal University took us under his wing for a couple days and drove us down to the ceremony at Shei-Pa National Park.

In the afternoon Dr. Lue hiked us along a ridge trail through a misty forest. The trees—dark, gnarled silhouettes in the fog—were mainly red cypress, an endangered Taiwan endemic in the same genus (*Chamaecyparis*) as eastern white cedar and Port Orford cedar, which grow along the east and west coasts of the United States. Where the trail approached the mountain edge, the wind blew a brief opening in the mist, and we glimpsed a vast expanse of ridges and valleys. Walking a few steps further, I noticed a promising looking rock near the trail—about the size of a large book—sitting on undisturbed ground in such a way that it could have held a hiding place beneath it. I lifted the rock.

Beneath the rock sat a salamander, gray and wide-mouthed, grinning up at us (fig. 6.2). She was about as thick as my thumb and speckled with light blue. If this had been the Berkshires, she could easily have been mistaken for one of the members of the unisex-

ual salamander complex. But this was the endangered Guanwu Formosan salamander, a Taiwan endemic with populations so small that Dr. Lue believes inbreeding was to blame for the asymmetrically shorter toes on her right foot.

After Taiwan Sydne and I stopped for a few days in the Netherlands before heading back to the United States. There, a Dutch herpetologist, Richard Struijk, graciously dragged me in the pouring

Figure 6.2. (a) Guanwu salamander in Taiwan, (b) unisexual salamander in Massachusetts, (c) eastern red-spotted newt in Massachusetts, and (d) common newt in the Netherlands.

rain through muddy ponds filled with common newts. People often ask me what newts are, assuming they are something related to, but different from, salamanders. Newts are definitely a distinctive group, but newts are indeed members of the salamander order. In fact, out of the ten families of salamanders, newts are members of the one family named Salamandridae, which includes the species *Salamandra salamandra*. (There is even a subspecies named *Salamandra salamandra salamandra*.) This is the type species after which the whole order of salamanders is named. It doesn't get any more salamandery than that.

Richard taught me how to spot the newt eggs, which were laid singly at the tip of underwater grass-like leaves. The giveaway was that the last half inch of the leaf had been folded over to protect the egg. I wondered if this insight might help me find eggs of newts back in the United States.

Later in the week Sydne and I walked through a Dutch outdoor sculpture garden whose paths were lined by beech trees. With their smooth, light gray bark and Crayola-green spring leaves, these trees felt no different from the beech trees of New England. The European beech in this sculpture garden, like the beeches of China and Japan, are in the same genus (*Fagus*) as American beech.

Beech, cypress, salamanders. These all share something in common. As Sydne and I bounced around the Southern Hemisphere during the prior five months, we hadn't seen any of them. Instead of salamanders, we saw side-necked turtles, a distinct evolutionary branch of turtles that occurs throughout the Southern Hemisphere and whose members hide their heads by folding their necks over to the side rather than retracting straight back like the storybook turtles of the North.

Instead of the "regular" beech we're used to, in Argentina we hiked through huge, dark montane forests of *southern* beech (*Nothofagus*)—trees in a different family than the northern beech (fig. 6.3). A few weeks later we hiked through similar forests of

southern beech in New Zealand. And, while our plane was crossing the southern Pacific Ocean, flying within sight of Antarctica, we read that paleoecologists have collected fossil pollen of southern beech trees from that ice-covered continent, remnants of the ancient Antarctic southern beech forests.

As Sydne and I experienced, Europe, Asia, and North America share species and higher taxonomic groups in common. Likewise, different sets of species are shared among the world's southern continents. Lots of taxa follow these same basic patterns. Why?

It goes back to plate tectonics. Today the Atlantic Ocean separates populations of European and American salamanders, allowing them to diverge along distinct evolutionary trajectories. But that hasn't always been the case. Europe, Asia, and North America all derived from the supercontinent of Laurasia. Back when these landmasses were connected, beech, cypress, and salamanders could move back and forth across the supercontinent. Even after the spreading North Atlantic Ocean finished separating Europe and Greenland fifty million years ago, occasional land bridges across the Bering Strait would have still allowed for some exchange.

Just as salamander ancestors casually strolled between Can-

Figure 6.3. (a) Southern beech forest in New Zealand, and (b) southern beech in Argentina.

ada and Europe in the past, an analogous mixing occurred among today's southern continents—who once formed the supercontinent of Gondwana. Five hundred million years ago—before salamanders had even evolved—Africa, Antarctica, South America, Australia, and New Zealand were all part of that gigantic landmass. Eventually, Gondwana would join up with Laurasia to form Pangea before everything splintered apart again. But even during the days of Pangea, when the landmasses were all united, there may have been an environmental barrier to north-south movement. The equator bisected Pangea, and the equatorial climate would have been markedly different from the climates further north or south. This likely deterred Gondwanan and Laurasian species from crossing over. Ultimately, it was across the unions forged by Gondwana that southern beech, side-necked turtles, and a whole variety of Southern Hemisphere specialties dispersed.

Of course, the world's biogeography isn't all so simply divided between north and south. There is, for instance, a distinctive set of snapping-turtle-sized trilobites whose fossils can be found along the margins of Europe, New England, eastern Canada, and northwest Africa. From these fossils geologists deduced that somewhere around 480 million years ago a chunk of Gondwana broke free and drifted off toward Laurasian landmasses, carrying its unique species along. This little Gondwanan chunk was the microcontinent of Avalonia.

And here I am, near the end of my morning road trip, in Avalonia. Within two hours I've gone from North America to Africa, from the land of salamanders to the land of side-necked turtles. It's true that land animals probably hadn't even evolved when Avalonia left Gondwana. And it's also true that 150 million years after Avalonia turned its back on Gondwana, all the world's landmasses reunited in the supercontinent Pangea, squishing Avalonia in the middle and allowing some species to disperse across the continental borders. But still, it feels biogeographically potent to now be

standing on a chunk of original Gondwana when essentially all of my life, save for a brief honeymoon, has been strictly Laurasian.

TO THE MARSH

At last I reach the water. Forgetting about rocks, this spot is a Mecca of its own, well worth the pilgrimage for any inhabitant of our valley (fig. 6.4). In rapid succession a cormorant, a Cooper's hawk, and a great blue heron fly past me over the water. A blue crab tiptoes across the bottom of the boat launch, partially obscured by the red surface reflections of my canoe that now floats in the water. Standing on shore, I give the canoe a hard push and jump onto the stern all in one motion. I sidle down into my seat as the boat and I drift away from the mainland.

Figure 6.5.
(a) Greater yellowlegs, and (b) tracks near where I landed my canoe in the marsh.

We cut across a channel as wide as a six-lane highway to a flat, grassy expanse on the other side. As the canoe and I approach, snails crawl across a flat mud shelf near the waterline, and I see tracks of a shorebird that I guess to be greater yellowlegs—they have the classic "game bird" foot morphology with three straight toes toward the front and one barely registered nub out the back (fig. 6.5).

Working our way a bit north, we find the mouth of a little meandering creek, about the width of a two-lane road. The tide is just barely beginning to come in, and this slight current helps move us up-creek. Bands of colors line the banks. Two-foot-high mud walls, dark brown, display exposed roots and a network of holes. On top of the mud walls stand erect grasses, three feet tall, dingy yellowish at the base and bright green on top. When your eye is at exactly the right level, a thin white stripe of salt cuts high across the grasses, marking the top of a very high tide.

We round a bend and find a tiny inlet off of the creek, just wide enough to squeeze the front of the canoe into. Behind the inlet is a long, straight channel lined with *Phragmites*—common reed. Fiddler crabs flee across the muddy banks, diving down holes as the canoe wedges in. I scramble up the steep mud slope, clinging to tall grasses and scattered logs to keep from sinking, wrestling to keep my boots from being sucked into the muck. Once I'm up the bank, the land is flat and firm. Next I drag the canoe up, hauling it a dozen yards over land and lashing it to a big piece of driftwood. High above me I catch the white-and-black markings of a flapping osprey.

It's a beautiful morning out here, despite the lack of salamanders. It's not the sun or the bedrock that keeps salamanders away; it's salt. Put most salamanders in salty water, and osmosis will suck water out of their bodies, right through their sensitive porous skin. Out of around 700 salamander species worldwide, only about a dozen have adaptations that allow them to tolerate brackish water—and they do so by raising their internal saltiness to compensate for the increased saltiness of their environment. But today the primary vertebrates here with me seem to be birds—the last lineage of the dinosaurs. A great egret, bright white in the morning sun, now skims across the grass, gliding downward for a landing (fig. 6.6).

As I wander through this flat expanse, patterns begin to emerge (figs. 6.7–6.9). The plants are not uniform. There are patches of red, green, yellow, brown, and tan. There are sections where the vegetation is tall, short, and even absent altogether. Some of the species seem to swirl together, like a van Gogh sky. Occasional straight lines cut long gashes through the canvas.

The burning question in my mind today is, "What causes these patterns?" I've come with two words in my head, "zonation" and "disturbance," but the theories hinted at by these pieces of jargon don't seem to fully capture what I see here (fig. 6.1).

The plants are not uniform, and they're also not random. For

Figure 6.6. Greater egrets and snowy egrets in the marsh.

the most part, I'd call them "clumped." Whether or not things are arranged in a random fashion tells us something about the processes at work—so it's useful to develop an eye for randomness. If you look at a little patch of the night sky, say, right around the North Star, the arrangement of stars is basically random. But if the patch of sky you're looking at includes the Milky Way, other galaxies, or star clusters, then you'll notice a clumpy distribution of stars. Like cows at watering holes, things that are clumped have some affinity for each other and so tend to cluster together.

If things tend to avoid each other, they will also be nonrandomly distributed, but in the opposite way from clumpy distributions. They will be evenly spaced, or overdispersed. Consider trees in a forest—trees don't want to grow too close to other trees, so they tend to be spaced just far enough apart to let their neighbors grow. On the bottom of a pond, frog tadpoles will often clump together as they munch down dense patches of algae. In the same pond, the carnivorous salamander larvae will be evenly spaced along the bottom, each defending her own little hunting ground.

Figure 6.7. Patterns in the marsh: (a) wavy lines, and (b) straight lines with changing vegetation height.

Figure 6.8. A muddy patch in the marsh.

Figure 6.9. Aerial view of the chapter site showing many patterns.

As an inlander, I don't really know much about coastal systems. Luckily, there are only a few dominant species here to keep track of: smooth cordgrass (*Spartina alterniflora*), salt marsh hay (*Spartina patens*), black rush (*Juncus gerardii*), marsh elder (*Iva frutescens*), and common reeds (*Phragmites australis*). These are the characteristic plants of this natural community—the salt marsh. Salt marshes form within the intertidal zone—where the daily fluctuations in water levels alternately submerge the plants and then abandon them to the drying air. And they only form on parts of the coast that are protected from the violent ocean waves.

I look up at the crescent moon in the blue sky and imagine a line connecting the endpoints of the crescent, extending down to the horizon. This is a navigation trick: where that imaginary line hits the horizon is, more or less, south. On the horizon south of here, barely visible in the chapter-opening image, is a thin line of elevated land far in the distance. This is Long Island, a big pile of debris—the terminal moraine—dumped at the end of the glacier that once sat here. As the glacier retreated, it left behind an island of rubble that now protects us here in the Long Island Sound from harsher ocean waves, helping this estuary to thrive.

More than just sheltered by the Sound, this salt marsh is protected from waves by the Connecticut River. We are just inside the mouth of the river, hiding in part behind a mound of fine sediments dumped by the river—the river's delta. The spot where we are standing is part of a much broader estuary—where fresh and salt water mix—extending up the river.

Besides protecting this spot from waves, the mound of sand may have saved the river from development. The shifting sands, continually renewed by the glacial sediments washing out of the Connecticut River Valley, historically made it tricky to navigate big boats into the mouth. So, unlike every other major river in the region, we didn't build a huge port city at the mouth. Instead the estuary was left to function largely intact. Today this whole estuary is officially listed under the international Ramsar Convention treaty as a "wetland of international importance," supporting critical breeding ground for many fish, birds, and other creatures.

Salt marshes are often lauded as the most productive ecosystem on Earth. Per square foot, these plants produce more biomass than anywhere else. I've heard this said a lot, but I've never fully understood it. I'm told it's because there are a lot of nutrients here. But why are there a lot of nutrients here? How did they get here? I need to go track down a biogeochemist who can spell it all out for me.

LUNCH

Hungry, and with many questions on my mind, I head for a huge log—a whole tree carcass that has washed in—to find a seat for lunch. As I walk toward my lunch destination, I wonder, could

Figure 6.10.
Predated crab.

that tree have once grown on Rainbow Beach before the under-cutting river dropped it and floated it all the way here? Eventually, anything—stick, stone, hat—plucked by the current up there would have to pass by here. Stepping forward through swirling salt hay, I notice for the second time today an empty broken carapace from a small crab sitting upside down in the grass, far from the water's edge (fig. 6.10).

Above me a monarch butterfly drifts along—one more in a steady drumbeat of monarchs that have been passing overhead all morning. One at a time, with long minutes in between, the orange flecks in the blue sky are all heading in the same south-southwest direction. Using their hardwired ability to tell compass directions from the sun, how many will make it to the wintering grounds 2,500 miles away in the forests of southern Mexico? Did any of these individuals flying over me start their lives in the milk-weed plants growing in my yard a hundred miles north? How do caterpillar-sized bites of leaves possibly contain enough energy to justify such an enormous journey?

Figure 6.11. Gull pellet with crab parts.

I reach the log and have a seat, trying to avoid the many white streaks and splats painted on by birds. Wedged into a crack in the wood is an irregular mass, about as big as the end of my thumb (fig. 6.11). Looking closely, the mass holds a mosaic of small, colored, sharp-edged fragments—blue, brown, pink, lavender, orange, gray, yellow—mixed with a few pieces of grass and coated by a shining layer of mucus. The fragments have the same speckled texture as the crab carapace I stepped over in the grass. In fact, these fragments may be the missing parts of that very same crab.

Earlier this morning a gull plucked the crab out of the shallow water and flew it over to the flat, low grass. There the gull flipped the crab and swallowed the choicest bits, including much of the underside shell and a few stray pieces of grass. The crab bits slipped down the gull's throat into the first part of her stomach, where digestive juices started breaking down the meat. Then the meal traveled to the gull's gizzard, where the sandpaper walls, assisted by small rocks stored just for this purpose, further pulverized the crab. But some of the hardest parts of the crab refused to be broken down. These hard fragments were compressed into a tight ball and pushed back up. The gull, standing on this log, pumped her neck, shook her head, and puked up this pellet.

Despite the prevalent belief that only owls make pellets, making pellets is a normal part of eating for most birds. Indeed, many other dinosaurs made pellets as well. It's just that owl pellets are particularly good for school children to dissect. Owls, having weak digestive juices and the tendency to eat their prey whole, often expel pellets from which you can reconstruct the entire prey skeleton. Plus, because of their size and charisma, owls make large pellets that are easy to find, easy to work with, and easy to get excited about.

PESTS

Perhaps the pellet was made by the same gull that now shifts nervously, eyeing me from the tall wooden nesting platform behind me. Similar platforms are scattered all across the marsh. They are intended not for gulls but for osprey. Osprey, the giant raptor whose wingspan can reach almost six feet. Osprey, the graceful flier who hovers in place with bent wings, high above the water until it spots a fish, then dives in for the kill. Osprey, the specialized hunter who sinks her talons into the meal, gripping on with uniquely barbed foot pads and reversible toes, then angles the load aerodynamically forward as she thrusts the still-wriggling fish through the air. Osprey, the globally abundant species that very nearly went extinct.

It was in this very spot where, sixty years ago, Roger Tory Peterson first documented the collapse of the osprey populations. Spurred on by the work of Rachel Carson, Peterson and his disciples ultimately traced the osprey decline to the popular insecticide for which Paul Hermann Müller won the 1948 Nobel Prize. In 1935 Müller started his search for the perfect weapon against insects. In 1939, after trying 349 failed chemicals, he finally placed a fly in a cage with DDT. The fly died. DDT went on to international fame, reducing populations of pest insects and helping in the fight against diseases like yellow fever and malaria. Crops, livestock, people, houses, and wetlands worldwide were all dusted with DDT.

Here on this estuary, as on many other estuaries, DDT was sprayed indiscriminately from low-flying planes in the middle of the twentieth century. The target was the salt marsh mosquito, whose larvae are specially adapted to live in the salty standing water that collects in small pools in tidal marshes. This aggressive mosquito bites both day and night, carries diseases such as Eastern Equine Encephalitis, and can fly up to forty miles from her breeding ground to badger inlanders who never set foot near the coast.

DDT worked wonders on the salt marsh mosquito. And globally, DDT worked wonders on many other mosquitos. By 1961 DDT had been used to eradicate malaria from thirty-seven countries, including from the United States.

But the DDT miracle turned out to be short-lived. In large tropical countries abounding in mosquitos, a tiny fraction of the mosquitos happened to have genetic mutations that allowed them to survive encounters with DDT. While most mosquitos crumpled and died around them, these mutant mosquitos thrived and multiplied. In evolutionary terms DDT exerted a new selective pressure, and the mosquito populations adapted. Having lost much of its potency, DDT could no longer be relied on as the ultimate cure for the disease epidemic. It was the overuse of DDT that caused its loss of potency—had we been more selective about which mosquitos we targeted for spraying, resistance would have evolved much more slowly in mosquitos, and perhaps we would have had better long-term success in eradicating malaria.

Though DDT's effectiveness waned, the chemical compound itself persisted. It clung to the small particles of clay and sand in the soil. It worked its way into the sediment layers. It was consumed by the microscopic plankton floating in the water column. Inside the plankton, DDT clung to the creatures' fat, even as other food by-products were expelled. Filter-feeding clams and fish consumed these DDT-laced plankton. Again, DDT clung to the fatty tissues of the clams and fish, becoming ten times more concentrated as it refused to let go. The DDT then took another step up the food chain when bigger predatory fish and omnivorous crabs feasted. Once again the DDT became even more highly concentrated.

At last seagulls and ospreys swept down and ate the crabs and big fish. These top predators, with concentrations of DDT hundreds of times higher than the concentrations of DDT in plankton, suffered. Their cognitive functions declined. Their immune systems declined. Their reproductive abilities declined. And, most obvious to human observers, their eggshells thinned to the point that they cracked too early.

In 1940, 400 ospreys nested here in the Connecticut River estuary. By 1970 there were only 16. DDT and the equally toxic compounds that it breaks down into, DDE and DDD, can last for many decades out in the marsh, poisoning not just the birds but all the big fish and mammals high on the food chain. Erecting nesting platforms, conducting scientific studies, and lobbying politicians to ban DDT, a handful of ecologists like Roger Tory Peterson, Rachel Carson, and Paul Spitzer fought to save the osprey. And they succeeded.

By 2010 there were once again over 450 ospreys breeding here. Staring at these nest platforms in front of me, I wonder, did Roger Tory Peterson—inventor of the field guide, leader of the twentieth-century environmental awakening, muse of the global birdwatching movement, Presidential Medal of Freedom recipient, two-time Nobel Peace Prize nominee, and self-described "hermit who lives up in the woods"—personally erect these very ones?

It's important to keep in mind that Rachel Carson and other en-

vironmental pioneers never sought a complete ban of DDT—nor
has it ever been completely banned. Rather, the point is to use it
wisely and selectively so that its potency is maintained. Where a
serious public health risk exists, most agree that DDT should be
used. But instead of spraying it across every cornfield, residential
garden, pasture, and swamp, spray it inside houses and at targeted
breeding sites in malaria-prone areas. And instead of relying on
DDT as the only answer and a magic solution, look to the other
tools in our toolkit. Use a variety of insecticides, consider new
technologies like the genetic engineering of mosquitos and fungi,
and rely on old-fashioned approaches, like eliminating standing
water where mosquitos breed.

FARMING

Long before DDT was popularized, people were fighting to
eliminate mosquito habitat from this salt marsh. Regularly spaced
ditches, dug in long lines throughout the marsh, allow fish to
swim up and eat the mosquito larvae living in marsh-top pools.
More to the point, these ditches act like reverse irrigation to drain
the water out of the pools so there's nowhere for the mosquitos
to breed. Normally, at low tide, water stuck in the middle of the
marsh would have to percolate sideways through hundreds of feet
of marsh to get back out to the open water. That's a slow process
and would hardly begin before the tide came back in again and refilled the pools. But if every spot on the marsh is at most a few feet
away from an open trench, the water rapidly drains away. The water table drops, and pools no longer sit on the surface. Mosquito
larvae dry out. But so do the plants.

Although the vast majority of the ditches in northeastern salt
marshes were dug as a part of the economic stimulus policies in

the 1930s, nominally for the control of mosquitos but really just to create jobs, some parts of these marshes had been cut with ditches for hundreds of years. And it wasn't all about mosquitos. Hay made of various salt marsh grasses—salt hay—is a prized commodity. Livestock love the salty taste, and seventeenth-century colonists would graze their cattle on salt marshes to cure sick animals. Farmers learned early that draining water from the marsh helped improve conditions for growing, harvesting, and grazing of salt hay. Today, more than 90 percent of the marshes in southern New England are striped with ditches.

If I'm really going to talk about the salt hay economy, I don't want to just read and write about it. So I recently ventured out to my local garden store and asked for a bale of salt hay. The old gardener behind the counter informed me that their salt hay comes from Boston's North Shore, where it's not a National Seashore, and so farming there is still legal.

Salt hay has a high value, the gardener told me, because it lasts years longer than regular hay before it rots, and, most significantly, it won't introduce weeds to the garden you're mulching. All bales of hay contain errant seeds from the grasses, clovers, asters, and other plants swept up in the bale. But the plants in the salt hay bale require salty conditions to sprout and thrive. The various salt grasses, the sea lavender, the seaside goldenrod, and the seaside aster pose no threat to my inland garden.

A burly staff member walked me out back to the shed where a forklift held a few of the treasured bales. Little dark seeds protruded from the bundle, and I recognized them from the salt marsh. Priced at $14.99, $3.00 more than a regular bale of hay, I bought one. Driving away, the sun in my car warmed up the bale, releasing a thick briny smell into the air. At home Juno and Alder were thrilled by their new present, which they decorated with flags, pretended was a horse, then smelled long and hard to see if they could pick up the scent of the coast.

I finish up my lunch on the log, but I'm in no hurry to move. I arrived at low tide, and now I really want to see high tide, which isn't until 3:00 PM. I've seen the marsh at its driest, and now I want the tide to roll in and soak me and everything around me, lifting my canoe off of the dry land. To have this experience, there's nothing left to do but wait.

It strikes me that as I've revisited my field sites for this book I've always been in a hurry. I'm working on a deadline to capture images and write a story before the seasons slip past, all while finishing up languishing academic papers and attending to the needs of my two children. If I sit back and take my time to do anything, a swarm of guilt, like a million salt marsh mosquitos just waiting for me to stand still, will bear down on me.

It's the typical pace of modern life. I raced out to Bark Hollow. I squeezed in quick visits to Maggie's Forest. I ran through the Great Swamp. I paddled hard by Rainbow Beach. Having been to each place before, I already knew the patterns I'd come to document, and I just needed to snap some pictures and get out of there. If I'd been living life at such a pace years ago when I first visited those sites, I never could have found the stories in the first place. To learn the unexpected, you've got to sit back, open your eyes, and let the patterns talk to you. I wonder how many new patterns I missed this year as I raced through the sites? Wasn't this whole project really just an excuse to experience nature at the pace I enjoy?

Here the tides are out of my control, so the guilt doesn't land on me. Happy at blaming the moon, I pull out my ukulele and strum a few chords. Hours pass, more birds and butterflies fly over, but not much else happens. High tide should be coming soon, but still the water doesn't seem to be anywhere near me.

My mind wanders to the crescent moon that was peeking though the trees last night before I tucked the kids in. And then it hits me. In my haste to assemble my gear and find a window in my life long enough to slip down to the coast, my planning had been negligent. I should have taken the kids on that moonlight wander in the woods after all. Sitting here on this log, my feet aren't going to get wet today.

LIFE EXPLODES

I get up off the log and wander over to the edge of the ditch to see if the water is still rising. I poke some tiny sticks in the mud at the waterline and then wait a few minutes. The sticks are now an inch down from the waterline. Indeed, the water is still creeping, ever so slowly, up into the marsh.

Although not much has changed back by the log, there are definitely big changes over here by the ditch. All the creatures have come to life. I see fish chasing blue crabs under water. A tiny flat crab crawls under my boot. Ghost shrimp race back and forth (fig. 6.12). The fiddler crab holes are now completely under water. Moon jellies drift by.

As I witness this explosion of life, it's clear that salt marshes really are extremely productive. The banks of the ditch seethe with crustaceans. I try to form the analogy between this marsh and the upland forests I'm used to. Instead of earthworms and millipedes on the forest floor, it's crabs and shrimp. Instead of oaks and pines, it's cordgrass and reeds. Is this place really so different from Bark Hollow?

Well, at Bark Hollow the plants are duking it out over calcium and magnesium. As the mountain rocks break down, these elements and a host of other macro- and micronutrients infuse into the forest soil where the roots fight over them. But almost as fast as the nutrients emerge from the rocks, the spiteful rain pours down and whisks away the calcium, magnesium, and anything else it

Figure 6.12.
Ghost shrimp.

can carry. Full of nutrients, water seeps through the ground, tumbles along tiny surface rivulets, creeps into streams, cascades into pools, and rushes into a small river that at last joins a large river. And where does that river lead? Here.

It's quite the pilgrimage that each nutrient atom has taken to get here from the uplands to the sea. Tumbling down the waterways, an atom of phosphorus spirals between two lives. Sometimes she floats freely as a dissolved phosphate, rushing down with the current. Then some organism gobbles her up, and our phosphorus atom lives for a while bound in nucleic acid, ATP, or another molecule. As she is passed from critter to critter, she doesn't move far downstream. Then she's excreted, or her host dies alone, and she returns to a free-floating life, headed for the sea before being captured by another creature. She's not alone. Nitrogen, carbon, sulfur, calcium, magnesium, potassium, iron, chlorine, and many more nutrients spiral downstream alongside her.

As I stand in this marsh, these nutrients swirl around me at the edge of the ocean. The ocean is the repository for all that's leached from upland soils. That's why it's so salty. And that, it suddenly hits me, may be why the salt marsh is so productive. The roots have access to the vast stores of nutrients washed down to the ocean—not to mention the nutrients deposited with the river-borne muck trapped by the tangled marsh. These are the nutrients that water stole from all the mountains, hills, and valleys in the other chapters of this book.

As I watch ghost shrimp float up and down in the ditch, I'm aware of the three-dimensionality that water provides life. Along every submerged stem in the marsh is a fleet of microscopic organisms—algae, bacteria, fungi, protists, viruses, and decaying matter—the periphyton. This layer forms a little factory that efficiently recycles local wastes and captures nutrients as they float by. This supports the plants and provides nutrients up and down the stems in a way that wouldn't be possible along the stem of a plant on land.

Why aren't land plants surrounded by such active living coats? For one thing there aren't as many nutrients floating in air as there are in water. But mainly it's because if these periphyton organisms tried living on the stem of a sunflower, they'd dry out. Of course, *water*. The most important nutrient. Second only to shelter on the list of survival priorities. It's needed for photosynthesis. It's the way we move molecules inside our cells and throughout our body. It's the basis of all life.

Sure, calcium and magnesium help determine which plant species grow on which patch of land, but they don't really have a big effect on the overall productivity of the land. Water, more than all those other nutrients, is the factor that most limits the growth of plants on land. Water itself is the reason marshes are so productive.

At last I think I get it. Of course this is one of the most productive ecosystems on Earth—it's rich with water and mineral nutrients stolen from the uplands. I guess I won't need to track down a biogeochemist after all. But I still need to come back here—under a different moon.

TIDAL FORCES

I look over to the part of the marsh where Olivia choreographs the other *Field Naturalist* students in an impromptu dance about the tides. One student is the moon, one student is the sun, and one student is Earth. Oh, and it's Halloween, so the moon and Earth happen to be wearing appropriately colored animal suits—tiger orange moon and blue bunny Earth. The outstretched arms of the Earth-student reaching toward the moon are meant to be the tidal swell in the oceans. As the Earth-student spins on his axis, his arms stay aimed at the moon-student slowly tracing her own orbit.

The yellow sun-student also exerts a tidal pull on the Earth-student, though only about one-third as strong as the moon's. When the students are standing in a line, the tidal arms reaching toward the moon are also aimed right at the sun. This is when the pull is the strongest, and thus the tides are the highest. At these times the moon is backlit by the sun, and Earth sees a new moon. When you look to the sky and see the moon near the sun, looking like just a little sliver—which my seven-year-old self decided was the "snag" of the fingernail I'd thrown out the window on our family road trip—tides will be extreme.

This all seems intuitive—that every time the moon passes overhead, a high tide will follow as the moon pulls the waters upward. But that's only half the story: generally, there are two high tides every day, not one. One of these high tides is on the opposite side of Earth from the moon. You can picture two bulges in the oceans on opposite sides of Earth circling daily. The Earth-student really needs to have his two arms always projecting in opposite directions from each other—one to the moon, one away.

Though the moon's gravity pulling water up toward it seems easy to believe, I find it hard to believe that there's a tidal force pushing water up on the opposite side of Earth. Professor Kannan Jagannathan recently set me straight on this, declaring, with a grin, that "tidal forces are real, whereas gravitational force is not real," at least according to Einstein's theory of relativity.

To understand tides I traveled back to my old college physics department. I was aiming for 10:00 AM, to catch an old tradition wherein the department faculty and staff come together to sip coffee and trade stories. The first thing I noticed was that the smells hadn't changed. I cut through the newer biology building, still packed with the scent of modern finishing materials that I had expected to fade years ago. (Is it the linseed oil in the linoleum flooring, some glue in the wood laminate, or some other composite material?) Then I passed into Merrill, the old brick behemoth that houses physics. Merrill smells of stone dust from the walls and greasy vapors emanating from a massive underground mechanical room whose door is often ajar, inviting mischievous undergrads like me to sneak through like a stowaway on a military submarine.

Morning coffee is held in the machine shop, where I once fashioned aluminum screws and clamps to help measure the electric dipole moment of the electron. I opened the door to the machine shop and was overpowered by its familiar oily smell. I found the machinist, Dan Krause, alone with the coffee. When I was an undergrad twenty years ago, Dan had already been there for decades. Now retired, his red hair had turned white and thin, but he still had the familiar habit of smoothing his hair by slowly running his two hands in synchrony from his forehead over his scalp and down the back of his head.

I asked Dan why no one else was at coffee. With his hands he mimicked texting on an imaginary iPhone and said that everyone is now so busy communicating on their devices that no one wants to sit and chat. Dan then lamented that even the new machinist, though superb, tends to program instructions into a computer rather than manually steer the metal lathes. In the sea of old machines, one smell not still in the air—in part because I'd stayed up until 4:00 AM cleaning it off of every surface in a corner of the lab—was the exploded beer bottle I'd tried to cool with liquid nitrogen one evening.

At last I found the open office door of Professor Jagannathan, affectionately called Jagu. In a red-and-white plaid shirt, Jagu's lanky frame was hunched over his little MacBook. I hadn't told anyone I was coming, and he sprang to life when he saw me.

Imagine three parts of Earth: the water in the ocean on the point closest to the moon, the solid body of Earth, and the water in the ocean on the point farthest from the moon. The water on top of Earth sloshes around, but Earth itself moves as one big rock. To simplify things, Earth itself can be represented by a point right at the center.

The force of the moon's gravity is strongest in the places that are closest to it. So the water closest to the moon is pulled more strongly toward the moon than Earth itself. This explains the high tide when the moon is overhead—all the water on the moon-side of Earth rushes and squeezes over trying to get closer to the moon and piling up underneath it.

On the far side of Earth, the moon is pulling more forcefully on Earth itself than on the water. In essence the water is staying still as the moon rips Earth further away from underneath. On the moons of Jupiter, this tidal ripping force is so strong that it powers massive volcanoes and earthquakes. Near black holes, it can shred stars into spaghetti.

But that explanation, as Jagu tells me, only works if you're someone like Isaac Newton who believes in gravity. If you are in an elevator, freely falling through a bottomless shaft, you wouldn't experience gravity. Hold out your left hand and drop a stone. It won't seem to fall. Instead, the stone just floats in front of you. You, the stone, and the elevator are all falling toward the center of Earth together. That's why Einstein says gravity isn't real: your experience of it depends on your perspective.

Still falling, now reach out your right hand and drop a second stone. Both of those stones are now falling toward the center of Earth. If you draw a line from each stone to the center of Earth, they won't be exactly parallel. There will be an ever-so-slight angle between them. As you and the stones fall together, look very carefully, and you will see that the stones will drift toward each other as they travel their separate paths.

This effect doesn't depend on your perspective. Whether you are standing on the thirteenth floor waiting for the elevator, or you are hurtling to your death with the stones, you will see that the stones are getting closer to each other. The difference in gravity experienced between two places—between the stone in my left hand and the stone in my right—is what Einstein calls the tidal force. In Einstein's theory of gravity, only these tidal forces exist, and they describe the geometry of space-time itself, which is warped in the presence of massive bodies like Earth and the moon.

So then. If you are Earth, holding the Atlantic Ocean in your left hand and the Indian Ocean in your right hand, with the moon high over the Atlantic, what do you feel? Every part of you is pulled toward the moon; however, because the moon is on your left, that hand is pulled moonward more strongly than your right. Floating freely in space, you don't actually feel the moon pulling your body at all, just as a skydiver doesn't feel Earth pulling her downward—she just falls. But as a celestial body holding an ocean in each of your hands, what you do feel is your left and right hands getting stretched apart, as if pulled in opposite directions. You feel the *difference*. The tidal force. The force that powers volcanoes on the moons of Jupiter. The piece of gravity that Einstein believed in. You feel your two handfuls of ocean pulled out away from the core of your body—and this is the reason that two tidal bulges rise up on either side of Earth. And this is the reason that each day on the marsh brings not one, but two high tides.

This is the simplified version of the tides. To be precise you have to add in the sun—whose tidal forces are weaker only because the sun is so much farther away from us than the moon, so the difference in gravity from one side of Earth to the other is much smaller. And beyond the sun, there are all the complexities of our shorelines, underwater topography, the latitudinal change in the moon's orbit around Earth, and other factors, like the lag between when the moon is overhead tugging at our waters and when the tides actually catch up to us many hours later. Travel the globe and you will find that these contributing forces vary widely from place to place.

When my friend Chelsea was working in Antarctica, she experienced just one high tide per day. When Sydne and I were rescuing whales in New Zealand's Golden Bay, I remember the tides rushing in like a fast-moving wall. In Canada's Bay of Fundy, daily tidal fluctuations can be over fifty feet because the moon's orbit matches the natural frequency of waves sloshing through the bay and connected waters. Like pushing a child on a swing, mother moon pushes child water through the swing bay with perfect timing to reach huge heights. Complexities abound. But in general, if you look up at the moon, you can make a reasonable guess about how the tides will behave.

THATCHED ROOF

After explaining the tides Jagu invited me to lunch with the other faculty, but I had to run off to meet Charley so we could teach an animal tracking course. As I walked back through the familiar smells in the physics hallways, I wondered what others thought of the smells emanating from me that last summer I worked in the lab. No one ever complained, but, that summer, I slept half a mile into the woods behind Charley's house in a shelter known by the Algonquian-derived word "wigwam"—a word that in some contexts has been used in a derogatory way but in its strict sense refers to a particular architectural form framed by saplings (fig. 6.13). I committed to a completely rustic sleeping experience—I left matches, flashlights, and all other electronics in the physics lab or in my car. I used friction to start all my fires under the cooking stone. I bathed only in the tiny stream nearby.

Often I would get out of the physics lab after dark and have to feel my way through the woods to the wigwam. One moonless night was so dark I couldn't see my hand in front of my face. The only light came from the occasional glowing fungus on branches lying on the ground. I crept along in bare feet, using my ears as much as my soles to follow the trail. Because the leaves on the trail were compacted, the sound of crunching was often the first sign that I'd made a wrong step. But the wigwam was about 200 feet off of the trail, on the far side of a small hill. I blindly abandoned the trail where it seemed right and stumbled through the forest.

After a long time away from the trail, I finally caught a whiff of the wigwam. Smokey, musty, unlike anything else in the forest. I put my nose in the air and walked upwind toward the scent. When the scent got weaker, I would turn toward where it was stronger. If you could see smells drifting off of an object, the plume would look like an expanding cone. Animals find the source of a smell by zig-zagging back and forth working their way up the cone.

Decades ago entomologist Jeff Boettner used to rear spongy moths in his laboratory, and he worked so closely with them that the spongy moth pheromones stuck to his clothes and got into his system. For many years afterward, Jeff smelled like a spongy moth, and free-flying spongy moths would seek him out. Standing in line at an ice cream stand one day, Jeff caught a glimpse of a lit-

Figure 6.13. Wigwam (sapling hut) insulated with common reeds.

tle moth on the other side of the street searching for a mate. The moth was downwind from Jeff. It flew first to the left, then to the right, slowly making progress upwind. He decided to have a little fun with the unsuspecting people around him. Beckoning with his finger, Jeff whistled loudly as if the moth was his dog, and he called out, "Come moth! Here mothy moth!" The people watched, wide-eyed, as the moth steadily advanced until it was right on top of him. Then he clapped his hands loudly and said, "Enough!" His hand claps mimicked the sound of a predatory bat and the frightened moth flitted away. Then Jeff called him back. "I'm sorry mothy, you can come back. Come back mothy!" And the moth came back. Again, Jeff clapped the moth away, and then again, he called the moth back. The slack-jawed ice cream eaters were stunned.

Working my way toward the smell of the wigwam that night, I didn't quite succeed. At some point the smell was so overpowering, I couldn't tell which direction it was coming from. Moth antennae are shaped like old-fashioned radio antennae and, like the independently smelling nostrils of dogs, are built for telling the direction that a smell comes from. My nose isn't. Exhausted, I lay down and slept in the leaves. A few hours later I woke up to a forest flooded by moonlight. There was the wigwam, fifteen feet in front of me.

A Woodsy Club project, the wigwam stood about eight feet tall and was about ten feet in diameter. Over a few weekends we bent saplings, peeled bark from a fallen basswood, and dug an underground vent to feed the fire. It was covered with a foot-thick shaggy layer that gave the wigwam an endearing look and that was the source of its distinctive smell. This gray-brown fur was harvested from a roadside wetland next to our local Target and, like traditional thatched roofs around the world, was made entirely of common reed, *Phragmites australis* (fig. 6.14).

Reeds act as an invasive species, and as we cleared several truckloads of the brittle stalks from the Target wetland, we hardly made a dent in the population. After we were done the wetland looked like the same dense monoculture of reeds as when we started. For the last 150 years, reeds have been invading wetlands across North America, altering the dynamics and crowding out native species. But common reeds are a native species. They are found on every continent except Antarctica and have been here in North America for at least 40,000 years—which is the age of Pleistocene sloth scats found full of reeds.

Although common reeds are native to North America, not all members of a species are the same. For instance, among rare California tiger salamander, a few invasive genes are spreading rapidly throughout the state. These genes, which originated from Texas salamanders imported for their quality as fishing bait, transform the salamander into an aggressive predator taking over wetlands and gobbling up other amphibians. Similarly, in common reed the recent takeover is blamed on the introduction of a European form

Figure 6.14. Common reeds invading an inland wetland.

of the species—although some botanists deem these as two distinct species. Before this introduction our native reeds were somewhat rare and grew at low densities mixed with other species. The invasive strain is stronger, taller, faster-growing, and more adaptable than our native strain. It has spread rapidly. When you see a wetland filled with reeds, it usually is a sign that a wetland is ecologically degraded, like in the Target parking lot. But very occasionally you find the native reed still meekly growing as it once did.

ZONATION

Lost in thoughts, I'm still standing out in the salt marsh waiting for high tide. In front of me is a cluster of what I think may be the native reed, mixed in with marsh elder. Compared to invasive reeds, these reeds are smaller, have redder stems, and grow more sparsely. This reed cluster is concentrated along the edge of the drainage ditch that runs down the center of our chapter-opening image—this is one of the many ditches dug for the purposes of mosquito control or hay farming.

What is this ditch doing to the marsh?

With our salt marshes so extensively assaulted by ditchdigging and hay farming, there's something really remarkable happening—or not happening. In most cases if you go dig up a forest or start farming a prairie, you're going to end up with a lot of species that weren't previously growing there. Some will be the crops themselves—wheat, corn, rye, potatoes. Others will be ubiquitous invaders taking advantage of the new early successional habitat—lamb's quarters, red clover, common plantain, dandelion, and so forth. But in these salt marshes that's not happening. Every plant that I find here seems to be a native species that's lived on this marsh for thousands of years. And in all, there really aren't that many different types of plants growing here.

In a way, the reason that there are so few plants here is the same reason that there are so many. That is, there are few *species* of plants, but many *individuals*—making for a very productive ecosystem with very little plant diversity. It's a pattern you see in many places. While increasing diversity often corresponds with higher productivity to a point, when you get to the extremes of productivity, sometimes you arrive back at something closer to a monoculture. There's a lot of debate among ecologists about when, whether, and why such patterns exist. But one explanation is that the richest environments are also the hardest to tolerate. In such conditions only a few species figure out how to survive, and those that do thrive. Here on the salt marsh, it's the abundance of water and nutrients that makes this place rich—and difficult to tolerate.

I like drinking water, but I can't survive more than a minute or so at the bottom of a swimming pool. Similarly, without special adaptations plants submersed under water will starve for oxygen. Additionally, the bacteria that grow without oxygen in waterlogged soil have a tendency to release chemicals like hydrogen

Figure 6.15.
Cordgrass, like many salt-adapted plants, excretes salt crystals through glands on its leaf.

sulfide, organic acids, and soluble forms of manganese and iron, all of which can build up under water to levels that are toxic to plants. And if the water then recedes and the soils dry out again, when these toxic chemicals react with oxygen in the air, they produce a whole new suite of toxic compounds that can further injure the plants.

I also like the electrolytes in Gatorade, but I wouldn't survive drinking only salt water. For a plant dunked in the ocean, the high concentration of sodium and chloride ions can wreak havoc. While chlorine is a micronutrient, needed in small amounts by plants, there's just way too much of it in the ocean (fig. 6.15). The osmotic pressure from the saltiness makes it hard for the roots to soak up water, despite being surrounded by water. Similarly, these ions will interfere with the uptake of other essential nutrients like nitrogen. What's more, the chlorine ions will enter the cells and break down basic metabolic processes by reacting with enzymes and proteins.

Thinking like a salamander, it's bad enough trying to imagine adapting to salty ocean water, but, given a few million years and enough motivation, we could probably figure out how to raise our internal concentration of salts to match the ocean environment. But now you want us to switch rapidly back and forth between environments? One minute we're deep in cold salty ocean water. Then the tide recedes and we're surrounded by brackish river water. Then suddenly we're exposed to the dry air and the hot sun. Any water that's left nearby starts to evaporate and the saltiness spikes. And then once again we're plunged back into deep water. Twice a day we're to handle the fluctuations in salinity, temperature, soil waterlogging, and changing current directions? No thank you.

And one of the most extreme things about northern salt marshes is the freeze-thaw cycle. Expanding ice crystals have the power to break rocks, push glacial till up through farm fields, and certainly to tear plant flesh. It might not be so bad if the ice formed once at the beginning of winter and then melted in spring. But twice a day the plants are dunked back in the water where the ice melts away, and then twice a day the plants are again exposed to the power of ice-crystal formation right down to their cores. These freeze-thaw cycles are why mangroves—the woody equivalent of salt marsh grasses—can't survive at high latitudes.

So back to my first question in this chapter: What creates the striking visual patterns of a salt marsh? If you Google it, you're bound to quickly encounter the word "zonation." Simply put, there are different ecological zones within a salt marsh, and different plants grow in different zones. There's the low marsh zone that is flooded every day by the rising and falling tides. There's the high marsh zone that is flooded only occasionally on extremely high tides. And then there are all the intermediate levels along this gradient. Moving from the lowest marsh up to the highest marsh in order, you pass through zones dominated by smooth cordgrass, salt marsh hay, black rush, and finally marsh elder.

At the core, zonation is caused by the fact that tides are harsh. Each species in the salt marsh has developed some level of tolerance to these harsh tidal conditions, and it doesn't like to be pushed beyond its limits. Imagine clearing out all the plants and then planting the species in the zones lower than the zones they belong in. Try planting salt marsh hay where cordgrass normally grows, black rush where salt marsh hay normally grows, and marsh elder where black rush normally grows. All the plants will die.

The unforgiving nature of the tidal zone explains why so few species grow here. It explains why, after being dug up and farmed, the marshes aren't covered with corn and clovers. And the specialization of these plants for particular tidal zones explains why a bale of salt hay at the garden store is worth a few extra bucks.

In our experiment with moving salt marsh plants around, now imagine moving the species to the zone higher than where they belong. Because they've now moved from a harsh environment to a gentler environment—at least from the perspective of terrestrial organisms—all the plants will live. If we plant cordgrass where salt marsh hay normally grows, it will grow just fine. That is, assuming we removed all the salt marsh hay first. The cordgrass *will* die if we just plant it in amid the salt marsh hay. The reason is that the cordgrass invests more of its energy in dealing with the harsh tidal environment and less of its energy competing with neighbors for space.

Less worried about the tides, the salt marsh hay, meanwhile, builds thick robust root systems that outcompete any cordgrass for the nutrients in the dirt. The same is true for each zone transition on up into the uplands: plants in the lower zone are more tolerant of harsh conditions, whereas plants in the higher zone are better at competing for nutrients.

Although there are a lot of nutrients here, they're not unlimited. In the uplands nutrients were less limiting simply because water was even more limiting. But here, with plenty of water, nutrients—particularly nitrogen—are generally the limiting factor for productivity.

So to add another twist to our experiment, imagine fertilizing the marsh so much that nutrients are no longer limited. Then the ability to compete with your roots won't matter. The most important limiting resource to compete for will be sunlight, and since cordgrass is taller than salt marsh hay, the cordgrass will invade and take over the salt marsh hay zone.

And what does the ditch do? It lowers the water levels, dries out the marsh, and shifts the zones. Cordgrass zones are invaded by salt marsh hay and then black rush. That's why ditching helps the farmers—since those latter two species are the more valuable crops.

Right at the edge of the ditch the effect is dramatic.

When turbulent tide water surges over the lip of the ditch, it slows down and drops any sediments it was carrying right at the ditch edge. This creates a little mound that runs parallel to the ditch. Often this natural levee is built on top of the mound of peat and mud that the Civilian Conservation Corps ditchdiggers of the 1930s dumped from their shovels as they dug. The result is a little spot of very high ground that supports high-marsh plant species at the edge of the ditch, even when the ditch runs through the middle of the low marsh.

And that's why we see a cluster of reeds and marsh elder at the edge of the ditch in the center of this site. Once established on the ditch edge, the reeds can then build the ground up even higher with their extensive root networks. Spreading clonally through sideways runners, the reeds will then creep down and send up new shoots in zones that would normally be intolerable. Still connected to life support from their parent shoots, these new shoots survive, building up their roots and raising the marsh height to a tolerable level, like that of the high marsh.

On this particular day the wimpy tides aren't getting anywhere near the high marsh. To see the whole of the marsh flooded, I'll need to come back in ten days—when the moon is full.

RETURN

Ten days later I'm back. This time Charley, Julia, and Juno are in the canoe with me. Within a minute of landing on the marsh, Julia spots a tiny speck on a glasswort. Glasswort is a funny succulent with stubby round limbs almost like those of a wee cactus. This season much of the glasswort is turning red and grows in little clumps here and there. The speck that Julia spotted on one of the glassworts looks like a miniature dried twig, smaller than a grain of rice (fig. 6.16). Charley and Julia spring into action photographing and collecting. Looking around, the glassworts are covered with these tiny grains of rice. Why all the excitement? They're the larvae of a case-bearing moth, a group of moths that build tiny houses out of silk and found debris. It is likely that this particular species hasn't been documented in the United States before.

Over the past few years, Charley and Julia have been photographing, collecting, and raising hundreds of leaf-mining insects, along with some stem miners. As larvae, these insects live inside of plants, digging tunnels through the flesh. Charley and Julia's home is overflowing with plastic bags and vials stuffed with leaves that house immature moths, flies, sawflies, and beetles. Some are in their refrigerator. Some are in their dark basement. After they dug up whole arrowwood plants to raise tiny, new-to-science bark-

Figure 6.16. Larva of a case-bearing moth that mines the leaves of glasswort.

mining moths, the four-foot-tall potted shrubs sat in their kitchen enclosed in hand-stitched sacks for when the moths emerged and started flying around. When they travel, their car is packed full of living specimens.

Leaf miners tend to be host-specific: each insect species prefers a particular plant species, genus, or family. But these creatures are poorly understood. Generally, if you find a leaf miner on an uncommon plant, there's a good chance it's an undescribed species. That's how Charley and Julia have discovered dozens of new species over the past few years. Want to find a new species? Go somewhere with uncommon plants and look for leaf mines, then plunk the leaves in a Ziploc bag and mail it to Charley and Julia to figure out.

Over the next hour on the salt marsh, in addition to the case-bearing moth, we find a leaf-mining moth in marsh elder that is almost certainly a species new to science, a stem-mining fly in salt-marsh aster that is either a new species or a species never before associated with this plant, a shore fly whose larval stage has never been described by scientists but seemed to be living on the aster Charley collected, and an aphid on sea lavender that Charley has seen in two other places and may be a new species.

As the four of us wander, admiring the patterns, the tide rises high. At last the tide washes over everything. It gets too deep for Juno's boots, and he starts walking around barefoot. It looks refreshing. Inspired, I do the same. Instead of sweating in boots and wool socks, now my feet feel the silky grasses under the cool water.

I point to a pair of dark, reddish lines that cut across the marsh, about five feet apart from each other. They are rows of glasswort. Charley and I debate the significance of the lines. Though Charley and I advertise ourselves as professional trackers, it's Julia who says they are the tracks from where someone drove across the marsh years ago. As soon as she says that, it seems so obvious: the pair of lines continue hundreds of feet into the distance.

So why is the glasswort growing in the tracks? The vehicle created low spots in the marsh—two long ruts in which water pools. At low tide the rest of the marsh is dry, but water still sits in the ruts, evaporating. Water vapor floats off into the air, but the salt stays behind, making the remaining water in the rut saltier and saltier. And because the soils remain waterlogged, the roots are continually starved for oxygen while the anaerobic bacteria are busily producing toxic compounds. Conditions in such areas become so toxic that only the most specialized salt marsh plants can survive, and often no plants at all. Glasswort is a specialist of such conditions. I wonder how many years these lines will last.

Of course, glasswort didn't evolve just for tire tracks (fig. 6.17). A host of natural disturbances can pave the way for glasswort. Somewhere on the marsh a raft of dead leaves from last year's grasses settles in a pile, smothering the vegetation. In another spot a block of winter ice freezes to the marsh, then a storm surge lifts the ice, ripping away the marsh surface. Elsewhere a voracious flock of geese tears up a patch. In all these patches the dominant marsh grasses are killed back, exposing bare dirt. There, glasswort seeds arrive, and the plant thrives until the clonal grasses creep back in and take over again.

When the grasses have been killed by a disturbance, their roots rot away, and this often leaves a hole in the marsh. Or, sometimes, the disturbance itself scours a hole. As in the tire tracks, water may pool here, and the soils may become waterlogged and toxic. In the center of the depression, the toxicity may be too much even for glasswort to handle. This leaves a large dead zone, which might become a permanent, self-sustaining feature on the marsh, known as a panne. Although the center of a panne might be devoid of vegetation, the edges are the most diverse places on the marsh—at least from a plant perspective. Sprinkled around the bare dirt is a sea of wildflowers. Red-stemmed glassworts are joined by purple flowers of saltmarsh false foxglove, pink flowers of sea milkwort, blue flowers of Carolina sea lavender, the tiny white flowers of seaside plantain, and several other species that grow in pannes and disturbed areas of the marsh.

Sometimes the levee formed at the side of a drainage ditch may ironically act to trap surface water right at the ditch edge, forming a little pool that supports panne species. Indeed, in the center of our site, a little red patch of glasswort suggests such a ditch-side formation. But pools of water, which might breed mosquitos and slow down hay balers, were exactly what the ditches were built to eliminate. And in general they succeed. The ditches crisscrossing this marsh have lowered the whole water table and mean that

high-diversity pools and pannes are much rarer here than they used to be. If you drive far up the coast into Maine, where fewer ditches were dug, you'll find many more pannes with much higher plant diversity. Perhaps as sea level rises, the water table in this southern New England marsh will rise too, and pannes will once again flourish.

Unfortunately, the more likely scenario is that the whole marsh will disintegrate in the rising seas. In this way the future might look a lot like the past; 20,000 years ago the salt marshes of New England didn't exist. How did they arise? Our marshes begin with cordgrass. At the leading edge of the marsh, cordgrass builds the marsh by trapping sediment in its roots and stems. Cordgrass is thus the engineer responsible for creating the habitat on which the other marsh species depend. Back when the continental glaciers were melting away, whenever pioneering cordgrass would begin to establish at the water's edge, the rapidly rising seas would push it back inland, leaving behind a marshless coastline. Then about 4,000 years ago, the rate of sea level rise slowed enough for cordgrass to keep pace. Ever since, cordgrass has flourished—expanding New England marshes upward and outward. In response to the slowly rising sea, cordgrass built up deep layers of peat beneath its roots. In this way cordgrass extended the area of the marsh inland as the rising waters slowly inundated the coast; simultaneously, cordgrass stretched the marsh out toward the sea with an expanding bed of peat. In areas where cordgrass was successful in building peat layers up to the high tide level, the drier conditions allowed salt marsh hay to then take over, adding on its own layers of dense peat. Today most of the marsh surface is dominated by salt marsh hay and other higher-zone species.

Salt marsh hay, with its cowlick swirls that typify our coastal marshes, now appears poised to vanish from this landscape. Marsh development all depended on the sea level rising at a slow enough rate for the grasses to keep up. With modern climate change, the pace of sea level rise is accelerating to rates not seen for thousands of years. In response, salt marsh hay zones are reverting to cordgrass, and cordgrass zones are washing into the ocean as the marshes simply disappear.

But compared to 20,000 years ago, the marshes do have one advantage: they're already here. Because the cordgrass doesn't have to start over from scratch, some scientists believe that certain well-positioned marshes may find a way to survive and even thrive under these new conditions by growing upward faster and migrating inland. That is, if human-erected walls don't get in the way of marsh progress.

Looking again at the patterns on the marsh, it's starting to make sense. Down by the water the plants suffer the insults of being dunked by every high tide and left dry by every low tide. But over by the log that the seagull and I ate lunch on, it's only the tides of the full and new moons that inundate the land. From low to high, there is an environmental gradient, and the plants echo this pattern. But the real marsh is not like cartoon drawings in a textbook, in which the marsh slopes evenly up from the sea. There are a lot of subtle ups and downs along the way. You might not notice one patch of the marsh is an inch or two higher than another patch, but that inch can make a big difference to the plants—it might be the difference between the roots being able to breathe air or the roots being drowned in saltwater.

Beyond simple gradients, the patterns on the marsh are reflecting past disturbances. Government-dug trenches. Tire tracks. Landing spots of marsh flotsam. I wonder how our footprints will be recorded in the marsh record.

But when I look at the solid swatches of color in the marsh, I still am not convinced that they all tell a specific story of elevation changes or past disturbance. I think there's a little more to consider. Moving from one patch to the next, the transition is sudden. While one foot is standing in a dense upright grove of cordgrass, my other foot is standing on the twisted wisps of salt marsh hay. As you move from one patch to the next, it's just like flipping on a light switch. If this were entirely controlled by an environmental gradient as you move uphill, wouldn't there be a gradual intermingling of species as you transition from one species' niche to the next?

If I may drag salamanders into this one more time, I'm reminded of a ludicrous set of salamanders that don't fit into any species concept—my favorite kind. For the past five million years, unisexual salamanders have been cloning themselves in ponds across the eastern United States. To reproduce, members of this all-female lineage steal sperm from one of five other species—whoever happens to be present in their pond—but then selectively discard most of the sperm DNA. The children, all of whom are daughters, inherit almost nothing from dad. These clonal salamanders dominate the ponds, crowding out the local host species by making near-exact copies of themselves.

Evolutionary biologists would usually say that clonal animals are at a big disadvantage. They don't get to enjoy the benefits of sex. Sex mixes up your genes each generation, allowing you to purge errors and invent new solutions to old problems. Without it you're doomed, and your lineage should be short-lived. But these crazy salamanders are incredibly successful. Their trick? Just a little bit of sex. Although they are mostly clonal, every once in a while a bit of dad's DNA leaks into the system, and that seems to be just enough for the salamanders to reap the benefits of sex without paying its hefty price.

What's the cost of sex? The most obvious answer is males. Males are a huge waste. A lineage free of males that clonally gives birth to only daughters can quickly outgrow a lineage that wastes half its energy on nonchildbearing males. And that's why this all-female lineage of salamanders dominates the ponds where old-fashioned sexual species still produce both males and females.

Beyond useless males, the whole process of sexual reproduction takes energy. Instead of shopping for the right mate, making eggs and sperm, and hoping to god that your offspring survive their early helpless phase, what if you could just grow a new copy of yourself whenever you felt like it by just sprouting a new limb? You'd be like the grasses in this marsh.

Sure, salt marsh hay can reproduce sexually, and it will do so just often enough to take advantage of the benefits of sex. In the meantime the hay multiplies by expanding clonally from its roots. Skipping the wasteful process of sex, salt marsh hay nurtures each new shoot through connected roots, giving it a leg up in life. In so doing, it crowds out all other plants that try to grow in the dirt within its slowly expanding patch.

And that, I think, is why the patches of color in the marsh are so distinctly separated. It's a wrestling match between clonally expanding patches duking it out over territory. Who grows where, in large part, depends on who got there first. It's not strictly about the soil chemistry and tidal elevation. In fact, as they creep outward, the plants at the edge of a patch may begin to grow in places they don't really belong. Sometimes, as with the common reeds expanding into the marsh from the ditch-side berm, the plants might reengineer this new spot to fit their needs—a self-fulfilling prophecy unfolding across the marsh.

MAJOR LESSONS FOR INTERPRETING A LANDSCAPE

- Consider the global biogeographic context of the species and rocks at your site.
- Look at the small-scale patterns of the elements of your site. Are they distributed in a random, regular, or clumped fashion? What is driving this patterning?
- Consider nutrient flows and chemical cycles at your site. What limits productivity?

(overleaf)
Figure 7.1. A lookout on patterns.

7. Elevation

As I hauled my canoe back up onto the mainland at the mouth of the Connecticut River, I met several locals who were out admiring the view. One older couple sat in a pair of lawn chairs, reading as the sun set over the marsh. A birdwatcher asked me what I was up to, and then he asked if I had read a recent book about canoeing the full length of the river, from the source to the sea. Starting at the mouth, if we were to swim upstream, the river would get narrower and narrower as tributaries branched off. Passing fork after fork of successively smaller rivers and then streams, we would eventually arrive at a tiny beaver pond on the border of Canada.

Floating the full length of the river is definitely on the list of things I want to do, although it always feels a bit arbitrary as to which of the many forks gets the honor of carrying the official name of the river. Is that beaver pond up by Canada so different from the beaver pond where the coyotes barfed-up voles at the head of another tributary of the river? If we really want to experience the full range of the river valley, perhaps we could seek out more ecologically meaningful extremes. Instead of traveling the length of the watershed from one end to the other, what if we were to go from top to bottom? At the lowest elevation we have a coastal salt marsh. And at the top? The alpine summit of Mount Washington in the White Mountains of New Hampshire.

I'd love to take the kids up Mount Washington, but a hike with toddlers to the top of the tallest mountain in the Northeast didn't seem in the cards this year. There is, however, a cog railway to the summit. It was getting late in the season, so we planned a trip for the following summer when the train would be running and the alpine wildflowers would be blooming. In the meantime I took an afternoon trip to a different summit in our watershed, Mount Cardigan.

By the time I reach the trailhead at 3:30 PM, the day feels almost over. Unlike all the other field sites in my book, I've never been here before. I arrived late because I was following my car GPS—it kept sending me on dirt roads marked by warning signs, "GPS route not recommended." Then the GPS batteries died. In the gravel parking lot, one other hiker is waiting to meet a friend so they can race up the mountain to catch sunset at the rocky summit.

Before entering the forest I spot some brown stalks of mullein standing in a little field and run over to collect them for our woodstove. It's been over a month since our new stove was installed in our house. The weather is getting cold, but I haven't let us start a fire in it yet. I want the first fire in the stove to be made the old-fashioned way: with fire by friction.

Cross-legged on the floor of our house, with the kids watching and mimicking, I've been trying and failing to make a fire by rubbing sticks together over the past few weeks. Why struggle at something so anachronistic? There's little natural about the factory-cast iron of the stove, and our house is loaded with matches, electric heaters, and propane burners. But as naturalist Jon Young once remarked to me, the woodstove is the heart of the house. And I want to hold something sacred in our heart. If I don't draw some lines for myself, I'd never be motivated to struggle. Sydne understands this line I've drawn and hasn't complained about the cold.

There are several ways to make a fire by friction, and the hand drill is my favorite. A set consists of two parts: a baseboard, which is a piece of wood about the shape of a TV remote; and a spindle, which is a pinky-thick, knee-high stick. You hold the spindle between your flat, prayer-like hands, rubbing palms back and forth to twist the spindle as you apply downward pressure. As you spin, smoke billows from the tip of the spindle and dust collects in a notch carved into the baseboard.

When things go right, the dust will be fine and black, and soon an orange ember will start to glow in its center. If the spindle is the male part, the baseboard the female part, the pressure from your hands the force of god, then the spark is the little embryo. Carefully, oh so carefully, you nudge the little ember into a fluffy nest-like womb of fibers you've prepared. You breathe into the nest, gently at first, then a little a bit harder and with increasing force, never pausing for too long so the fire doesn't collapse. At last the nest bursts into flames, and you transfer your baby fire into a pre-prepared home of small sticks.

With a good hand drill set made out of the right materials, when your body knows what to do, you can have flames in under a minute. Otherwise you can struggle until your hands are bleeding and you collapse from exhaustion.

In college when I was first learning to start fires, I would practice this daily. Bloody blisters on my palm near the base of my middle fingers turned to callouses. "Stigmata," my brother and I used to call these marks. I haven't made a hand drill fire in years, and my callouses are subtle. I still have a collection of old baseboards, the edges lined with dozens of little black circles from attempts at fire making. Looking at the circles, I can generally tell which ones successfully made fires—the notches are centered, the sides are straight, and the smell of smoky wood is perfect.

Once the muscles in my arms know what to do, there are really two keys to making fire. The first is mental centering. Trying to simply power through and just make fire never works for me. I can only make a fire if I first take my shoes off, sit up straight, close my eyes, do some meditative breathing, and say a little thanks to the wood, fire, and the elements of nature. I can't tell you how many times I've come to a hand drill kit saying to myself that all that meditation stuff is silly—it's just a physical challenge. I failed to make fire every one of those times. The other key is to have the right materials. For the baseboard, willow and tulip poplar have a nice soft grain—or you can go pick up a white cedar fence post from the hardware store. For the spindle, my favorite is mullein.

I stash the mullein in my car and start up the mountain. This isn't the highest mountain in the Northeast. Mike Jones suggested I come here because it's actually the lowest mountain around where you can find alpine specialist plants, particularly capitate sedge. Beneath an overcast sky the forest at the trailhead glows with fall color: orange leaves of sugar maple, red leaves of red maple, and yellow leaves of beech, white ash, yellow birch, and hobblebush. Dark green splotches of hemlock contrast with the bright colors.

I'm sticking to the trail on this hike, so I feel a bit lost on this unfamiliar mountain. A hermit thrush lands on the ground right in front of me, then flies up to the trail marker. A red squirrel takes a drink of water from a little stream, then leaps to a tree, scurries up, and chatters at me. The yellow blazes are hard to see against the fall colors. At a stream crossing I hear blue jays and a hairy woodpecker calling, and I see a dark wet stain on a gray rock where someone recently stepped from the stream with a wet shoe. I pass a cut through the trees intended for winter skiing and hear a bird alarm in the distance.

As I ascend, the forest changes. I encounter some big red maple and ash trees over an understory of beech. Further up, sugar maple completely replaces the red maple. Hemlocks fall away and instead there are just scattered spruces.

Racing against the dimming light, my right knee starts to hurt and I hear several red squirrels chattering. Up ahead I see a wall of dark green. Suddenly, I'm in a forest dominated by red spruce and balsam fir, with a bit of paper birch and mountain ash. The trees are small, thin, and dense. Whereas before the trail had been a soft bed of fallen leaves, now it's just bare rock with silvery flecks of mica glimmering through. Clumps of moss cling to the rocks. I hear the rapid thumping of grouse wings fleeing from my footsteps.

Then, just as suddenly, I'm back in the same deciduous forest of oranges and yellows as the leaf-covered trail continues to climb. I can see the mountain's bare summit through the trees. The sun pokes through clouds only to show that it's about to slip behind the mountain. Then I'm back in a forest of spruce and fir.

I pause at a lookout and take a picture of the view. I'm looking at the shoulder of the mountain, and there's a strange pattern. A dark green swath—shaped a bit like the Nike Swoosh—cuts across the orange mountain (fig. 7.1). What is this? I stare at the pattern trying to make sense of it. A few theories arise in my head. To test them I will have to return and walk through that forest in the distance.

DESCENDING

The sun sets, and I can see fog engulfing the barren mountaintop. I climb a bit further and reach a zone where it's mostly bare rock with scattered pockets of vegetation in little depressions here and there. The fog surrounds me, and it's ghostly quiet and still. Straining my ears, I hear what seems to be a distant saw-whet owl giving a series of shrill, rhythmic cries.

It's almost completely dark as I head down the mountain, and now it's my left knee that's hurting. Without a flashlight on I can just barely see the path. I assume I'll feel the difference in the trail beneath my feet if I step off the path, like when I used to walk to the wigwam after a long day's work in the physics lab.

Now it's really dark, and even the tiny red light of my voice recorder overwhelms the night. Am I still on the trail? I step intentionally to the side, and oh, yes, the sound of my feet on the leaf litter is completely different. It smells good here. I hear the brook bubbling and a distant barred owl. But mostly just a hush. It's moonless. Like that night I got completely lost on my way to the wigwam. At least tonight I won't have to leave the trail to get where I'm going. I can just listen to the sound of the trail and let my ears guide me down. Although there are lots more big creeks to cross on this journey.

I might be off the trail now, I don't know. I'm determined not to use my headlamp. But. It's going to take a lot longer in the dark. And. I have all these responsibilities. A life I have to get back to. Deadlines. Scheduling. Kids. Shoot, I need to make it back for their bedtime. If I were younger, I'd definitely go without a light. But I'm going to cheat a little just to make sure I'm still on the trail. As I make this decision and rummage through my gear for a light, I hear something bounding off through the woods.

I turn on the light, and I see the eyeshine of spiders reflecting back at me. The wash of light shows that I was going the right way. It was just self-doubt. Well, maybe I was a tiny bit off the trail, but mostly fine.

With the light on now, walking is much faster. But all I see is that spot. That circular spot of light. I've lost all awareness of the forest around me, of the trees, of my wide-angle vision. I'm no longer walking in the forest. I'm now walking in a spot, a circle, a cylinder—no, a little cone of light. I have no idea where I am again.

Noah, don't you remember that time we were walking at night in Black Rock Forest in New York and found a glowworm beetle? An amazing creature. The male has wings and giant antennae but is otherwise unremarkable. The wingless female looks like a glossy red, black, and yellow worm in the light. In the dark the female is a magical glow of faint blue dots. They're probably everywhere, but you never notice them because you have to walk around in the pitch dark for them to pop out.

I try turning my light back off. Right away I stumble. All I see is the afterimage of the spot in the center of my vision—even though I was using the red light setting. That darkness seems to suck me in.

I flip my light back on. Now that I've used the light, I'm addicted to it. The trail is actually easier here—wider and flatter. But I need this light more than I did on the steep narrow trail above. I'm moving too fast now. I'm rushing along in that headlights mode. If I want to turn this light off, I'll have to completely reset. Earlier I had transitioned slowly with the sun setting through the fog and the clouds. By the time dark came, I was ready for it. Now I'm hurried and have no time. I'm not listening, I'm not feeling, I'm not smelling. I'm just moving. I have to reset. I have to reset. I have to reset.

I stop and do a few meditative breaths. As I exhale, I feel my heartbeat and use that as a cue to center myself. OK, I can do this. I start forward. I trip. I'm still hurried. I turn my light back on. I'm just trying to do too much. I've got to get home to the kids and get ready for work tomorrow and all the other things I'm trying to cram into my life. If I'm trying to do all the things I'm trying to do, I can't slow down. If I can't slow down, I can't see the woods, I can't experience it, I can't be here, I can't feel it.

Those damn hikers I met at the trailhead going up to catch the sunset. They were the ones who reminded me to bring a headlamp. I hadn't even thought of the sun setting, nor had I packed a light in my bag. At their urging I went back to the car and got this headlamp. I wish I hadn't brought it. Then I would have been forced to experience these woods.

Crashing through with monster steps, I feel so out of place in this hushed forest. I should have my light off. I should be tiptoeing. I should be barefoot—that would force me to be here. As it is, I may as well be driving my car down this trail.

When the light's off, it's not that I don't have the ability to follow the trail. It's just fear. I'm afraid to move forward. What if I'm wrong about where I'm going? What if there's something dangerous out here? As if my narrow 10-degree cone of light allows me to see what's out here any better. As if whatever's out here will automatically be in my light, not in the other 350 degrees around me.

I hear my whole body pounding with every step. Thump. Thump. I hear it echo in the ground. Thump. Thump. I should slow down, but I'm still racing. Thump. Thump. I hear a mouse squeaking in the trees above. Thump. Thump. Writing this book gets me outside. Thump. Thump. But it's just one more reason to race. Thump. Thump. When all I really want to do is slow down. Thump. Thump.

SECOND VISIT

In the spring I return to Mount Cardigan, heading for the green swath that's been on my mind all winter. I have a simple quest to test my favorite theory for this pattern. Almost at my destination, I'm walking through a bright spring forest. A carpet of trout lilies, Indian cucumber, wild oats, wild sarsaparilla, and painted trillium. An understory of beech with a few scattered spruces. A canopy of red maple, sugar maple, beech, and yellow birch. A northern hardwood forest.

Suddenly, the forest shifts. It's dark. Beneath a spruce and fir canopy, I see lots of bedrock poking through the bed of needles. Moss and lichens grow on the ground amid a few scattered lowbush blueberries. A montane boreal forest.

Now, to answer the question. I dig two holes in the ground. In the hardwood forest the soil has a rich, pungent smell that reminds me a bit of the stink some millipedes give off. I dig and dig through orange dirt, little bits of debris wedging beneath my fingernails. I dig as deep as my bare hands are able and don't find what I'm looking for.

Over in the boreal forest, within a minute of digging through the thin soil, my fingers close on what feels like the tiny treasure I'm looking for. I pull it out, dust off the dirt, and inspect it up close. Yes! To be sure, I bring it over to a piece of birch bark to make a picture.

As the black flies swarm me, I'm happily content at having proven my theory. Well, I guess it's not proof, but supporting evidence. I mean, I've only dug two holes—I could have gotten lucky. If I were a real scientist, I'd need to dig many holes to make sure it wasn't dumb luck. But if I don't find what I'm looking for in the next holes, that could ruin my whole theory, and my book chapter. Better quit while I'm ahead, right?

The scientist in me says to dig at least two more holes. I begrudgingly agree. With enormous trepidation, I walk a distance and dig two more holes—one under hardwoods and one under the conifers. With my first scoop of leaf litter in the hardwood hole, I find a buried beechnut, possibly cached by a blue jay. I dig further but don't find what I'm looking for. In the conifer hole? Again I find the tiny treasures. Phew. I'm convinced. I'll stop here, leaving it to some young graduate student to dig hundreds of holes across the mountain for a full statistical analysis.

With plenty of daylight left, I head on up to the summit of Mount Cardigan, just over 3,000 feet in elevation. It's mostly an expanse of bare rock. Here and there are a few patches of vegetation clinging to crevices and sheltered spots. After a brief search in these green areas, I find the rare alpine specialist capitate sedge (fig. 7.2), along with other alpine plants like Bigelow's sedge, three-toothed cinquefoil, mountain cranberry, and crowberry.

The sun's about to set, and I sit and meditate for a bit. I stare out across the expansive landscape below. As I'm about to start down the mountain, I notice the slanting light illuminating long grooves through the bare rock. They point southward. Glacial striations. As high up as I am now, 20,000 years ago I would have still

Figure 7.2. Capitate sedge on Mount Cardigan.

been deep under the glacier. In its southward advance the glacier scoured these grooves into the bedrock (fig. 7.3).

I hike down without a light. The full moon is out. I'm more relaxed than last time, in part because my kids have become less dependent on me over these past few months. My body knows that they will be fine if I get home at 1:00 AM. I'm met on the trail by a couple wood frogs and an American toad enjoying the evening.

Suddenly, there's a little flash and a clanking sound. What was that? I take a few steps back, retracing my movements. I strike the metal tip of my hiking pole against the bedrock. That was the sound. I do it several more times, and then there's a spark. Wow. I do it again and again. I feel as ecstatic as a caveman first experi-

encing flint and steel. I set my digital SLR camera down in the dirt with the shutter open for a long exposure to capture the freshly exposed fragments of iron oxidizing in tiny flames as they hurdle through the air.

Back at the trailhead, I call out to a couple of talkative barred owls. The pond, which earlier in the day had bullfrogs, now has chorusing spring peepers. I get in the car and, as I drive along the network of dirt roads, I realize I've got no maps or GPS service. The moon is still out, and I picture it hanging in the south down over the salt marsh at the mouth of the Connecticut River. If I just head toward the moon, I should make it where I want to go. Until cell signal returns I turn onto the roads that aim toward the moon. I'm

Figure 7.3. Glacial striations on top of Mount Cardigan.

in no hurry. I still have a few hours left on my audio recording of *Zen and the Art of Motorcycle Maintenance*. So I drive off the mountain while two dueling personalities engage in a final showdown over the narrator's body.

CHASING TOADS

My knowledge of these mountains largely traces back to Mike and Liz, who, besides studying turtles, also have an affinity for the alpine. One particular evening sticks in my mind—camping with them up in the Lewis Hills of Newfoundland. After peering down into the 1,400-foot-deep canyon of Rope Cove, which led out to the Gulf of St. Lawrence, Mike, Liz, and I set up our tents near a small pond. I sat alone by the pond, wondering if we might find amphibians breeding at that elevation. The sun had just set over the canyon, trailed by a fingernail moon and Mars. I could still feel the long shadows of the barren red rocks, unable to distinguish the landscape where I sat from the images beamed to Earth by the Mars rover, except for that small pond. I sat in the spot where the clunky hooves of a moose passing at 4:00 AM the following morning would wake me for a staring contest.

Unexpectedly, my ear caught a trill that hung for a few seconds then vanished. An American toad. Or did I imagine it? Silence. I cocked my head and waited. There it was again, from north of the pond. It must have been real. I estimated that the toad was maybe a hundred feet away, probably on the far side of the pond, or possibly in a seep a bit uphill from there. I waited to hear the trill again, then stalked over toward the sound, taking my camera but leaving my GPS. Again in silence, I waited for another trill, then stalked toward it. Wait, trill, stalk. Wait, trill, stalk. I got to where the toad should have been, but the sound was coming from still

Figure 7.4. Arctic hare in the Lewis Hills of Newfoundland.

further upwind. The game continued as I made my way uphill, the toads never seeming closer.

I was carried further and further from the tents, but the stars were bright and good for navigating home. As I approached some tall grass, a low form broke the horizon nearby. It took a lazy hop and stopped to inquire about me. Gray and white spots of a shaggy milk cow, stout arms and legs, tall ears. This was my first encounter with an arctic hare (fig. 7.4). She seemed unconcerned, letting me approach to within ten feet. Then I backed away in peace to continue my toad quest. The toads had grown quite persistent and numerous by this point but seemed to be no closer.

Next I was interrupted by strange muted bursts, like a mischievous child playing with a creaking door in the grass below. It sounded like a pickerel frog to me. Though I couldn't find the culprit, I later learned that I was being mocked by a ptarmigan (fig. 7.5). Eventually, forty minutes after leaving our campsite, I arrived at a vast network of ponds on top of the hill, teaming with toads and

Figure 7.5. Rock ptarmigan takes flight in the Lewis Hills of Newfoundland.

other life. The word "moor" sprang into my head, and I expected the grave-haunting hound of the Baskervilles to leap into view.

The wind had carried the sound of the toads, and me, over half a mile. As if on cue from Arthur Conan Doyle, when I arrived a thick fog dropped down obscuring everything more than an arm's length away from my face, including the stars. If I headed toward home but veered a bit west of our tents, I might step off the vertical wall of Rope Cove. While I contemplated a life-threatening adventure back to our campsite, Mike and Liz appeared out of the fog on the edge of the moor, also drawn in by the toads.

WINDY SUMMIT

In mid-June Alder, Juno, Sydne, and I head to the Mount Washington Cog Railway, fifty miles north of Mount Cardigan. Before checking in to our hotel, we watch a video of a man unsuccessfully trying to eat cereal in gale-force winds on the summit. Now the kids are extra-excited for this adventure. We arrive at the train station in the morning and get out of the car surrounded by a bright green forest of birches, maples, aspen, and mountain ash.

It's a nice, sunny morning as we stroll into the ticket office for our prebooked trip up the cog rail. Suddenly, the ticket lady introduces unexpected turbulence into our leisurely morning. The first problem is that it's too windy at the top of the mountain for the train to go all the way. We will only be going partway up, and we won't be allowed to walk around and photograph plants when the train stops. Do we want a refund so that we can make the two-day journey to this mountain some other time? I desperately want to get to the summit, the kids desperately want to get on the train, and the price of the tickets is painful to think about. The second problem is that I had the time wrong. Instead of leaving in a half hour, the train is leaving right now, and we need to make a decision fast.

I decide to get a refund, since the trip won't be worth it. The kids, meanwhile, are out the door bounding excitedly toward the train without tickets. Beyond them, the train tracks climb steeply up the mountainside. There beside the tracks I see the pattern I expected to see back at Mount Cardigan. As the tracks grow skinny in the distance, the bright green broadleaf forest transitions into the dark green conifer forest before the whole scene disappears in a cloud. Rather than the sharp line I saw at Mount Cardigan, the transition here is gradual, with scattered spruce and fir trees at the base becoming slowly more abundant (fig. 7.6).

Why is the Cardigan transition so sharp? A few months ago Charlie Cogbill—who knows decades more about these forests than I ever will—told me that such abrupt transitions are pretty typical. There's a sort of feedback loop, wherein the trees dominating a stand shape the environment in such a way that promotes those of its kind. Once a site begins to be dominated by spruce and fir, it doesn't linger in an intermediate state—it quickly becomes completely coniferous. Maybe, but I don't fully understand. This slope in front of me seems to be telling a different story.

I turn back to the ticketing agent. OK, just give us the tickets and we'll get on the train. We stumble onto the sold-out train car,

Figure 7.6. Vegetation transitions up the face of Mount Washington.

like a little tornado family. Sydne and I have big packs for carrying children and lots of gear for the hike I thought we'd take at the summit, Juno's carrying his own overstuffed bag, we're flustered about the rapid decision, grumpy because there are no seats together nor window seats left, and the kids are bouncing up and down with enthusiasm. The sixty other tourists on the train, wearing casual Sunday clothes, totally unencumbered by gear, quietly stare while Alder and I squeeze in next to a family from Charleston, South Carolina.

The track is so steep, almost 40 percent grade in places, that the train uses toothed wheels that mesh into a toothed rail to climb. The steam engine fires up, and we begin to clamber toward where the track disappears in the clouds.

We cross the highest elevation tributary in the watershed of the Connecticut River, and the conductor announces that, ecologically, our three-mile journey up the mountain will be equivalent to something like a 600-mile trip north (figs. 7.7–7.10). When you climb a mountain, like when you drive north, it gets colder. In the case of going north, the curvature of Earth means that the ground gets less direct sunlight to warm each square foot. In the case of going up a mountain, the thinning air makes it colder.

Out the train window, spruce and fir slowly replace the birch and maple. Historically, these conifers were the prized timber in these mountains, and loggers working at the lower slopes would often cut out all the spruce and fir while leaving behind the maple and birch. Compared to the hardwoods below, the spruce and fir now surrounding us do better on shallow, rocky, acidic soils—acidity that their fallen needles help create.

(left)
Figure 7.7. Mount Washington broadleaf forest.

(right)
Figure 7.8. Mount Washington boreal forest.

Figure 7.9. Windswept trees on Mount Washington.

If we were on a road trip north from the eastern deciduous forest, at this point we would be entering the boreal forest, also known as taiga. Charlie Cogbill would be quick to point out that there's a huge difference between spruce-fir forests defined by elevation and those defined by latitude. My casually conflating a montane red spruce forest with a boreal black or white spruce forest makes him shudder with disappointment. Once I get to know these systems better, I'll probably look back and shudder alongside him.

The boreal forest, which wraps around the top of the world from Canada to northern Europe to Russia, is the largest terrestrial biome on Earth (fig. 7.11). The broad, flimsy leaves of maples

Figure 7.10. Alpine tundra on Mount Washington.

and oaks have a tough time with the long, cold winters. Conifers do much better here, with waxy needles designed to withstand freezing and drought and a conical tree shape that helps shed snow. Unlike hardwood forests, these stands of conifers burn vigorously, and so fire is an important component of the boreal ecosystem. Travel to the boreal forests of Canada and you'll find cold-adapted mammals like woodland caribou, lynx, marten, snowshoe hare, and red squirrel (fig. 7.12). These replace more southern species like bobcat, fisher, cottontail, and gray squirrel.

As the cog railway reaches its steepest point, the conductor invites everyone to try standing straight up inside the climbing train—which requires leaning far forward. Out the window the

Figure 7.11. *Field Naturalist* student in spruce-fir forest.

trees shrink. They are no longer big and tree-like. They are now only a few feet high, gnarled and twisted. *Krummholz*. That's German for "twisted wood." These trees have been pounded into submission by the wind.

A little black spruce forms a bud on its highest branch, aimed up to the bright sky above, hoping to outcompete its neighbors for light. Then a cloud sets down on the mountain with a strong freezing wind. Tiny water droplets from the cloud begin to freeze on the little spruce bud. The ice formations grow bigger and bigger until, snap! The wind breaks off the ice, along with the little bud. Instead of growing upward, the spruce turns to another bud, smartly tucked along its side, to send off a horizontal branch.

We pass beyond the steepest part of the ascent and, coming up over a shoulder of the mountain, the tracks begin to level out a bit. Now we're too high for even the smallest tree to grow. It's just a low green carpet among boulders.

The vast open landscape reminds me of Iceland—that little

country up at the Arctic Circle. Iceland was the last stop on our trip around the world, and we spent just a few days gawking at puffins high on the cliffs. But it was mid-May, and Sydne and I really wanted to see the winter Northern Lights. So we booked a short trip to Iceland that January and returned with our cross-country skis hoping for a relaxing week of skiing under the aurora.

When we landed back in Iceland, we instantly felt silly for having lugged along our bulky skis, which barely fit in the rental car. We'd come from New England where there was a foot of snow on the ground, we hadn't seen the buried grass for weeks, and we didn't expect to see grass for many more weeks. We'd gone skiing

Figure 7.12. Red squirrel replaces the more familiar gray squirrel as you move north into coniferous forests.

almost every day that winter, and now here we were far north in Iceland—where there was no snow.

Yes, Iceland sits at the Arctic Circle. But it's also an island in the middle of the ocean. Water, with its great thermal mass, modulates extreme temperatures. Ocean air keeps the winters from being cold. That particular ocean happens to be extra warm due to the warm North Atlantic Current. When we told folks in Iceland about our New England winters, they were horrified at how cold it sounded. Our weather blows in from the west across a big, cold continent with no ocean to moderate it.

Though we didn't ski in Iceland, we walked on black volcanic sand beaches, gazed in wonder at swirling green auroras, and enjoyed the vast treeless landscape. This treeless landscape, like the arctic climate, was also misleading. According to historical accounts, Iceland originally supported extensive woodlands. But logging for farms and fuel, combined with grazing livestock, denuded the landscape. With the trees and vegetation gone, Iceland's thin volcanic soils washed away. Even though many farms have been abandoned, without any soils and with few seeds, the trees have yet to regrow.

On the train up Mount Washington, we've reached treeline at around 5,000 feet. On our imagined 600-mile road trip, we'd be entering the arctic tundra—perhaps in northern Labrador, although we would have run out of roads there. Permafrost in the soils excludes roots; freezing temperatures cause ice crystals in sap to kill trees; and the brief warmth in summer isn't long enough for woody tissue to grow. In the northern tundra, arctic foxes chase ptarmigan and arctic hares through dwarfed shrubs, grass, moss, and lichen.

But the analogy between driving north and going up the mountain only goes so far. It's more than just cold that prevents trees from growing on top of this mountain. The wind, the ice, and the extreme mountain weather patterns are really what hold back the trees. Where they can find shelter from the winds, like in a crack of a big rock, some trees actually can survive up above treeline in the mountain cold.

The train pauses, engulfed by a thick fog, as the conductor talks into his radio. Then cheering erupts as he announces that we've been cleared to go to the summit. The train lurches upward. The ceiling is now undulating intensely from the strong winds above.

When the train reaches the summit, in preparation for going outside, we bundle up and I hoist my pack of gear and picnic lunches for the adventure ahead. What alpine flowers are blooming now? The Appalachian Trail (AT) crosses this summit, and Juno is excited after having hiked tiny bits of the trail in other spots up and down its length.

The first time I encountered the AT in these mountains I was in college. It was 2001, just a couple days after September 11. My *Plant Ecology* professor had taken us up into the mountains to experience how plants change along an elevation gradient. We hiked up a side trail along a ravine to a ridge where we encountered the AT.

I remember looking out from the AT down onto the vast national forest on the other side and falling in love with the view. At our feet were little red mountain cranberries clinging beneath tiny dark green leaves—the source of the lingonberry juice served in IKEA—which we popped in our mouths. Then a scraggly man with a huge reddish beard came hiking up the AT from the south. He'd been hiking since Georgia, 2,000 miles away. Had he even heard about the terrorist attacks yet? Did he know that the whole world was in a state of shock? Should we tell him?

As we exit the train, the conductor warns us to hold tight to bags and children in the wind. Outside we see what he means. It's around 40°F, and the winds are holding at around sixty miles per hour, with much higher gusts. It's enough to topple us over if we don't lean just right. One woman is actually blown over on the tracks. While the woman flails, we drag the children over the tracks toward the shelter of the summit house, fighting to stay on our feet the whole time. Hiking now seems absurd. Juno and I enjoy a couple brief ventures into the wind. He's ecstatic about this weather, but it's too much for Alder.

After a brief stay the train gathers its travelers and heads back down the mountain. Juno and I now sit together, and we talk about the plants as we descend back down through krummholz, spruce-fir forest, and the forest mixed with maples and birches. Then Juno turns and delivers a lecture on steam trains to the passengers behind us.

Eating our peanut-butter-and-jelly sandwiches amid the sunny 70°F day back at the base of the mountain, we decide we didn't have enough time to enjoy the alpine plants. So we drive over to the other side of the mountain, where a paved road offers another route to the summit. We climb back up into the wind, stopping periodically to search for miniature alpine plants.

Once out of the car, Juno and I jump off of rocks to let the forceful gusts carry us far through the air. We pour water from our bottles toward our mouths only to have it ripped off the mountain before reaching our lips. Finally, Alder decides he wants to go play in the wind too, proclaiming simultaneously that it's "too windy" and that he wants "more wind!"

We soon find three-toothed cinquefoil and Bigelow's sedge, like on top of Mount Cardigan. Then Juno and I huddle over a pincushion plant, *Diapensia*, and remark that the flowers are bigger than the leaves. A little fist-sized cluster of tiny leaves clings to the ground beneath showy white and yellow flowers. A couple feet away a few woody-stemmed plants creeping across a rock, so flat that they're hard to distinguish from the nearby lichen, support finger-thick red spikes and gray fuzzy tufts. The shape of the flowers, I suggest to Juno, is a lot like those on the big weeping willow in our yard. Same genus, but this is bearberry willow.

These dwarf plants are uniquely adapted to the alpine. The

Figure 7.13.
Tolmie's saxifrage,
an alpine specialist,
on Mount Rainier in
Washington State.

growth form is so compact—either as mats or pincushions—that the plants seem to form a solid body, like a little animal, not unlike the stockily adapted form of the arctic hare. The leaves together create impenetrable skin protecting an interior stuffed with dead leaves and other organic matter. Each leaf itself is thick, waxy, miniature, and sometimes fuzzy. Many of the plants also grow deep taproots and turn dark red in spring and fall. These features protect them from drying winds and help them warm in the sunlight (fig. 7.13).

The diminutive size of alpine plants is about more than just hiding from the cold air. The soil below is also cold. It's so cold that the soil microorganisms become sluggish. These are the decomposers responsible for recycling essential plant nutrients. With a lazy recycling crew, dead organic matter isn't broken down quickly—it just sits around and piles up, trapping nutrients inside. But alpine plants are used to this. Perpetually starved for nutrients, they are adapted to stay tiny, so as not to outgrow their soil resources.

Here atop Mount Washington is a vast expanse of this alpine community, but it's all above 5,000 feet in elevation. The farther north we go, these plants can be found at lower and lower elevations, until, eventually, the alpine tundra merges with the arctic tundra all the way down to sea level. So why is it that fifty miles south of here we find the same alpine plants on top of Mount Cardigan at an elevation of only 3,000 feet?

On our drive back down the mountain, we stop in the krummholz, where the kids hunker in the shelter of the twisted shrubs, picking cones off of mountain alder. Near the base, back in the boreal forest, we go for a leisurely stroll under balsam fir and red spruce along another section of the Appalachian Trail, relieved at the windless quiet.

GALÁPAGOS

In any mountainous region, elevation gradients are a driving force on plant patterns. Here in the Northeast, valleys are lush, while mountaintops are barren and hostile. But in desert areas it's often the reverse. I've been studying Galápagos giant tortoises for the past few years, although, like a true modern ecologist, I haven't actually been to the Galápagos. I've just been modeling data on my computer, trying to predict how climate change will alter future vegetation and turtles on the islands.

It's a remarkable landscape down on those volcanic islands—or so the data says. The group I'm working with, led by James Gibbs, is particularly interested in the seasonal migrations of the tortoises between the highlands and lowlands. Up in the highlands, cool air and moist clouds produce lush greenery year round. Down at sea level the natural ecosystem is a cactus-dominated desert, with bursts of vegetation following the rainy season.

At least, Galápagos lowlands should be a cactus-dominated desert. But following the decimation of the tortoise population by explorers, the cacti, too, disappeared from much of the landscape. The cacti, it turns out, rely on tortoises to disperse their seeds—the cacti being a primary food source for the tortoises. Without tortoises woody shrubs have taken over where the cacti once were. Without cacti modern conservationists are having trouble repopulating the islands with tortoises. It's a real pickle.

In late May James came to Harvard Forest, where I am based, to present his work on Galápagos tortoises. After the seminar we went for a walk around the Forest. On the way out to the big experiments in the woods, we passed an old red pine plantation. It's a common sight around here—strange stands of uniform red pines all lined up like soldiers, with sparse vegetation below.

Neat and tidy. To me that means impoverished. But I've always had a ramshackle aesthetic and, like Pig-Pen, have always been surrounded by a cloud of entropy. I justify my disorder by claiming that ecological diversity thrives when things are messy. And the same can be said for ideas. Neatness is a hindrance to creativity.

Without the hand of foresters, such red pine forests would never grow here, and yet they're all around. What's funny is that, in fact, there is a native red pine community that does grow naturally around here. But native red pine is a community I rarely see, and it's usually protected as a special place. In this region red pine only sprouts well on exposed ridgelines along with other specialist plants—not in low, flat, shady forests like this one.

James and I wandered out to Harvard Forest's soil-warming experiment. There, underground wires have been heating the soil for the last twenty-seven years to simulate global warming. Digging through the soil, researchers like microbiologist Serita Frey have found that warming causes complex changes in the microscopic bacteria and fungi living underground. It's easy to think that a simple gradient from cold to warm, like descending a mountain, would cause a straightforward increase in microbial activity and decomposition. But it's much more complex than that, with many interactions among different species of plants and microbes, and globally relevant impacts beyond just nutrient availability. Which microbes live in the soil determines the rate at which wood and other organic materials decay. This, in turn, affects how much carbon dioxide the soil releases back into the air—in other words, how the soil breathes. But it turns out that twenty-seven years isn't long enough for the soil communities to reach a new equilibrium, so the experiment runs on, day and night.

As James and I walked back from the soil-warming experiment, we paused to look at a tall, lone pine with a stripe running down its side. Harvard Forest sits up on a hill, and we get a lot of lightning strikes, which frequently mess with the experiments. The scar on this tree showed black around its edges a few feet off the ground. It seemed to be an old scar, and James wondered out loud whether the black was really charcoal or if it was actually a fungus. Sometimes black fungus on wood can produce a remarkable imitation of charcoal. One way to tell is to try to draw with it. If it makes a black mark on paper, it's charcoal. If not, it's fungus. I leaned over and broke off a piece of the wood. Several attempts to draw on bark and my pants ended in failure. It must be fungus.

On our walk I told James about the book I'm writing. One of the chapters is about a barren rocky mountaintop that supports alpine plants at an elevation much lower than expected (fig. 7.14). On the flanks of the mountain, bands of spruce-fir forest slice through the hardwoods. Underneath the spruce and fir, bedrock juts up through thin soils, whereas thick soils hide the rocks of

Figure 7.14. Summit of Mount Cardigan.

the neighboring hardwood forests. The little black treasures in the soil explain it all. When I discovered the treasures, I then carried them over to birch bark to make a picture, literally. Dragging them across the white bark, they produced rich black lines. Charcoal. Charcoal that had been waiting for me in the soil for 160 years (fig. 7.15).

I don't know whether it started with a lightning strike, the flick of a match, a spark of steel, or a glowing ember beneath a turning spindle of mullein, but in 1855 a massive fire burned the summit of Mount Cardigan, denuding it of trees. With no roots to stabilize the ground, the soil then washed off of the summit, exposing

Figure 7.15.
Charcoal dug from the soil on Mount Cardigan, used to make line drawing on birch bark.

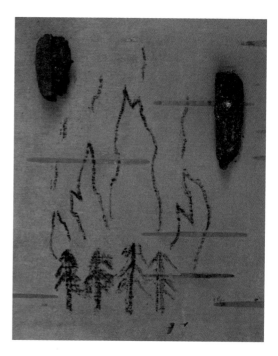

the bedrock below. Like in Iceland, the loss of soil changed the nature of this mountain, sending a closed subalpine forest back toward the first stages of primary succession. The top of Cardigan, devoid of protective vegetation, was now more like Mount Washington—exposed to the extremes of wind and temperature. It became a place where, despite being relatively low in elevation, only alpine specialist plants could survive the harsh conditions. And it's all because of that fire.

How exactly capitate sedge and the other alpine specialists arrived at this mountain still isn't clear. Maybe the summit had burned earlier and had been open for longer than we think. Maybe some seeds got stuck to mountain-hopping bird feet. Or maybe some mischievous botanist, seeing the alpine conditions, transplanted the plants for fun.

The fires that swept the summit evidently also burned down the sides of the mountain into the forests below, where the little black treasures lay in the soil. That explains the mysterious Nike Swoosh of dark green. It's a scar. Freshly burned soil makes a perfect seedbed for the conifers. Spruce and fir moved in where the old maples, beeches, and the soil beneath them were destroyed by fire.

But, we might ask, why did the fire burn where it did? Random chance? Perhaps. Or perhaps the fire is in turn a relic of centuries-old logging practices. Loggers, extracting the valuable conifers, may have artificially converted sections of the forest to hardwood-only stands. Where the weary loggers didn't reach, the conifers still stood. The hardwood stands resist fire. But the remaining conifers would have burned hot and fast, laying down a patch of burnt land that was even more favorable for future conifers and future fires.

MAJOR LESSONS FOR INTERPRETING A LANDSCAPE

- What environmental variables (e.g., elevation, moisture, temperature) define your ecoregion or ecological zone? What are the neighboring zones, and how do the species assemblages change?
- What past events—like fires, landslides, or mining operations—have left scars on your landscape?

8. Disturbance

GOING UP

I feel as exhausted as Sisyphus pushing his giant boulder up the hill. My hill is a 700-foot gain in elevation up part of the small mountain behind our house. My boulder is a double stroller holding two children. The path we're on is a washed-out dirt carriage trail from the 1800s, deeply rutted in places by flowing water and definitely not intended for strollers.

This week has seen our first set of really hard frosts. Icicles cling to the moss-covered cliffs on the trailside, and ice is heaving out of the ground like winter flowers. Sitting side by side, the kids are wrapped individually in bulky layers and then together in my sleeping bag. Alder, not yet talking, is just along for the ride. But Juno, now three, is ecstatic that we'll be visiting this particular field site. He's been here a few times before; this place is a topic of frequent conversation, and a three-foot-wide slice of one of the hemlocks from here—a "tree cookie"—sits on the floor of our living room.

We pass a few other hikers who shoot various looks of amusement, incredulity, and the typical patronizing approval for my taking care of tiny children alone—as if I couldn't possibly be their primary caregiver and as if being a nurturing dad couldn't possibly come naturally. Settle down, Noah, these aren't just the usual arrows of sexism: this outing is truly a bit extreme. A few raised eyebrows can be forgiven today.

We reach a particularly steep section of the road, and I'm not sure we can make it any further. Rocks are rolling downhill as my feet lose their grip and the three of us begin slipping backward. I pause. I catch my breath. Up ahead, the road levels off a bit and turns right—west. At the bend a little hiking trail drops off down along a steep cliff to the left—east.

That side trail ahead tumbles down massive formations made of impressive conglomerate rocks alongside a waterfall crashing to the base of the mountain. Below, the land is suddenly flat, and the creek gently saunters away from the pool beneath the waterfall. A few steps east of the waterfall there's a railroad, stretching long and straight along a north-south line. Cross the tracks, climb a short hill, and you find yourself under a high-voltage power line staring at low outcrops of metamorphic bedrock.

Two decades ago, miles south of here, the daily Amtrak, using that same set of tracks, would race past my dorm window on its way from Burlington, Vermont, to Washington, DC. One hundred and fifty years before, the waterfall was a popular railroad stop. At its peak of popularity in the late 1800s, vacationers would climb up the waterfall to take horse-drawn carriages to a resort hotel on the summit. But the hotel soon burned down. As legend has it, it was a band of local women who started the fire, upset that it was closer to a brothel than a hotel.

But the train tracks aren't there simply because of the tourist attractions—they follow a natural topographic line. This is the Eastern Border Fault. Standing at the power lines east of the tracks, we'd be on the Bronson Hill Island Arc, the volcanic islands that crashed into our continent 450 million years ago. But here on the west side of the tracks, we're in native North America. For today we're going to stay on this continent, as our site is even farther west and even farther uphill.

I turn the stroller around and dig in my heels to drag the kids backward up the slope. The stroller inches upward as my butt falls to the ground. I scoot my heels toward my body, gathering the strength to stand up. On my feet again, I use the weight of my body as it falls against the hill to heave the stroller a few more inches up. Again and again. In this way, standing, heaving, and falling uphill, we creep upward, following the carriage horses. The kids are having a blast.

Twenty minutes later we pass the spot where this fall I came across a dead porcupine belly-up in the trail. A few yellow leaves were stuck to the quills in its side. A couple feet away more quills were pinned into the leaf litter on the ground. Brown, peanut-sized scats were scattered about. The scene was completed by oak branches, each with pencil-thick bases pruned at a sharp angle and attached green leaves nibbled at the tips. Standing over the porcupine on my way back down the mountain, I saw a woman dressed in black hiking up toward me. She carried what seemed to be a piece of blank paper, ripped from a journal, as if she was going to the top of the mountain to write a poem or a letter to a friend. I waited for her at the porcupine so that we could console each other.

"Who'd we lose here?"

"Porcupine," I responded, and, having put the pieces together, added, "Fell out of the tree while eating."

"Life is challenging and uncertain," she offered. Silently, she continued on up the trail and I down.

At last Juno, Alder, and I arrive. The road continues to the summit, but that's for another day. On the south side of the trail, we're looking at a steep slope, much too steep for a stroller—even I can see that. The children are cold, hungry, and tired, and for these last 200 feet, they want me to carry them up the slope, together. To the dismay of my vertebrae, I oblige. With Alder strapped to my front,

Figure 8.2. *Field Naturalist* students at north part of site.

Juno slung over my right shoulder, my camera hanging at my left hip, and the rest of our gear in the torn pack on my back, we stagger up. Then we collapse at the base of a large tree, and I pull out lunch.

Giant hemlocks watch over us as we eat (fig. 8.2). It's dark down here. We're in a steep ravine between two hills, cut by a tiny intermittent stream. Scattered hardwoods—sugar maple, black birch, beech—add occasional splashes of color with their bright orange and yellow fall leaves. The ground is strewn with thick trunks of fallen trees. Tucked beneath some broken branches, Juno finds a bright red partridgeberry—pea sized—attached to a small green creeping plant. He picks it and we all share the tiny fruit.

I point to a stump up the hill and tell Juno, "That's where our tree cookie came from." I first came here twelve years ago, taking *Forest Ecology* from Bill Patterson, who among other things specializes in prescribed burns like those around Maggie's Forest. Bill had been studying this spot for decades. He told our class about several big chestnut oaks that were mixed in with the hemlocks near the top of the slope when he first arrived. But from 1979 to 1981 a huge outbreak of the invasive spongy moth killed off those oaks.

A massive hemlock had been blown down the summer before Bill took our class on this field trip, and when we arrived the tree was lying with its crown uphill, to the south. The department's tool-wielding Dan Pepin came out with his chainsaw so that we could inspect the tree's growth rings. Dan started the first cut into the prostrate trunk about fifteen feet up from the base. As soon as the chainsaw was through the wood, the weight of the roots suddenly tilted the whole fifteen-foot section back up again. Dan then had to cut it back down so that we could see the rings near the bottom of the tree.

I feel a deep connection to that fallen tree because I knew the storm that blew it down. The morning it happened I had been sitting inside my old house—a tiny rental cabin—petting my cat, Annie, trying to motivate myself to go outside. A short walk through the woods down the hill from my house was a great big wetland—a giant fen. I had made a resolution to myself that I was going to go out there and try to just live off the land for a week or so. I had cleared my schedule that week, and that was the day I was supposed to leave.

But it was one of those summers where everything seemed to turn upside down. The Crocodile Hunter had been stabbed in the heart by a stingray, I had been dumped after a five-year relationship, and my landlords had burst into my house while I was gone one evening, then left me an unsettling phone message telling me how to use a broom and threatening to evict me if I didn't do my dishes. By the end of the summer, to re-center myself, my sister had made me sit out in my yard making meditative rock spirals à la Andy Goldsworthy.

I was sitting with Annie on a little red loveseat that I had once carried on my back from a used furniture sale, gathering the energy to go out to the fen. In my bedroom a mouse that Annie had killed and lost was rotting somewhere in a pile of clothes. Pale, inch-long botfly larvae that the mouse had inhaled as tiny eggs wriggled in its flesh. The larvae emerged through the skin, crawled over to the carpet, and formed bullet-sized black pupae, which I later collected and raised into adult flies.

Suddenly, Annie embodied the look of a wildcat poised for action, momentarily frozen with wide eyes, perked ears, and tension coiled into all of her muscles. I heard it too. Was something banging on the metal roof? Hail maybe? No, the sharp pops were farther away. Snap, pop, crackle, and then, it was a screaming train.

I've heard it said that tornados sound like trains, but this sound wasn't just *similar* to a train—there seemed to be an actual train now racing through the forest behind my house. Always having wanted to see a tornado, I burst out the door to look. I couldn't see much through the trees, and then the train was gone. When quiet returned, I excitedly, and foolishly, ran out into the forest to look for the new train tracks. But my path was blocked. The popping that had given the train a definitive chugga-chugga sound seemed to have been made by whole trees splintering.

Most trees in the tornado's path—about 300 yards across—had been destroyed. If only I'd been out in the wide-open fen, looking up at the expansive view of the hill, I would have seen a tornado go by with my own eyes. I suppose I might have been killed too.

The tornado had cut a path three miles long, going right through the center of town—an eccentric, 250-year-old rural town with a proud motto: "We're all here because we're not all there."

I made my way into town amid the din of chainsaws just in time to see the sole victim being dragged out of a collapsed barn. Although she had a sprained ankle, the goat eventually made a full recovery. A half mile away I found what seemed to be a large post from the barn that had been picked up by the storm and planted upright in the center of a field.

For the rest of the summer, I made frequent pilgrimages to the path of the tornado, mainly attracted by the plethora of exposed roots and sloughing bark that I used to make baskets. The swirling winds had toppled trees in all directions. They were like handfuls of toothpicks that Alder had tossed on the carpet. Some trees were still standing, with their tops twisted off.

In the forest where Juno, Alder, and I now sit, it wasn't the tornado itself that pulled down the hemlock but, I believe, another isolated burst of wind generated in that same storm. And now a slice of this tree sits in our house, where we occasionally count its rings. I lay a piece of tape across the surface, and Juno helps me count: one, two, three. . . . When we get to ten, I draw a short line on the tape with black marker, and we count the next ten rings. In this way we march back in time through the decades: ten, twenty, thirty. . . .

DARK AND LIGHT

As Juno, Alder, and I finish up our snack amid the hemlocks, the sun slips down behind the hill to our west. I really want to climb to the top of the ravine. But even if they had the energy to walk, or if I had the physical strength to carry them, there's not enough light or warmth left in the day.

Besides, I was just here yesterday, alone. Though homesick for the kids, I was able to reach the other half of this site (fig. 8.3). I climbed up the dark hemlock ravine. Almost near the top, I found

Figure 8.3. *Field Naturalist* student at south part of site.

a dark brown scat—smooth-sided, blunt-ended—dropped by a bobcat. I reached the crest, then walked a few more yards, descending just a bit down the other side in time to capture photos for this chapter's opening image before the sun set.

Between the two sides, the forest couldn't be more different. Dark versus light. Tall versus short. Hemlock versus oak-hickory. Surrounded by ravine walls on all sides, like sitting in a dank basement, versus admiring the view through the trees that carries on for miles across hills and sky.

There in the bright forest on the other side, the *Field Naturalist* students are huddled in groups, still discussing the mysteries I'd set them on four years ago. The obvious question is, "Why is the bright forest so different from the dark forest?" But there are also a lot of smaller questions to stretch their minds (see fig. 8.1).

The students divide into pairs and rotate through stations. Over here an oval of moss, three feet by two feet, sticks up through the leaf litter. Why? Here a small sapling stands, dead. What happened to it? Here is a large rotting log. What stories does it tell? Here is a tangled branch on the ground. Why is it here? Off over there, where the slope of the hill turns away from us, are some understory hemlocks. What's going on with them?

HOME FOREST

While the students inspect, ponder, and debate, I lean back into the mountain, pondering the fate of the creatures that this mountain supports. On the other side of the mountain, on a north-facing slope a mile up the hill behind our house, sits the place we call The Crevice: two thirty-foot-tall blocks of conglomerate that face each other. In the winter the nooks surrounding The Crevice are home to porcupines and endangered hibernating snakes. In summer the snakes wander widely, some feasting on voles in our overgrown yard. But as fall sets in, they converge back at The Crevice. With the right timing, we can arrive to find several jet-black silky lines as thick as my wrist and longer than I am tall mingling on a carpet of bright yellow fallen birch leaves. All winter long the snakes huddle in a ball underground keeping each other warm. Without this hibernaculum they would die.

The snakes and porcupines were likely at The Crevice when Sydne and I hiked up there for our wedding ceremony—along with my ninety-year-old great aunt balancing a pumpkin on her head and Sydne's nine-year-old nephew collecting hundreds of partridgeberries to eat. The creatures were at The Crevice when we hiked up through the foot and a half of snow in the hours before

Juno was born—although not when we hiked up through the heat of summer for Alder's birth. And the snakes were certainly there in the dead of winter when we tossed the boys' placentas in. In all these hikes, from the silty lake-bottom ecology to a landscape littered with glacial boulders and stone walls, we followed the trails. But how do the snakes find their way?

While we are limited mainly to visual cues, many species of birds, turtles, snakes, salamanders, invertebrates, and others navigate by sensing Earth's magnetic field, by seeing the polarization of light in the sky, and by smelling cues in the ground. Twenty years ago researchers collected a bunch of newts from a pond. Then they put them in dark buckets surrounded by powerful magnets and drove them away in a truck, spinning them continuously. At the end of the ride, the researchers set the newts down in a laboratory twenty-five miles away from the pond. There the newts turned their noses to face the direction of home. Somehow each newt, without knowing how she got there, could sense where she was on the map in her head.

Although the snakes just hole up and sleep at The Crevice, the porcupines are after more than shelter here. Snakes don't need to eat all winter. Porcupines, on the other hand, do.

Just like the snakes, you can find porcupines foraging far and wide in fields and forests in the summer. Porcupines are perhaps my favorite local mammal. These gentle vegetarians keep to themselves, care deeply for their children, and are relatively easy to get to know. One summer after college I made friends with a local porcupine. Out on the lawn of the house we rented, the porcupine and I would lounge together, a few feet away from each other, while she picked up clover flowers with her front paws and munched them down. But to some people, porcupines are just big nasty rodents that chew things up. One day late that summer, my roommate looked out the kitchen window and saw our landlord hacking the porcupine to bits with a shovel.

As winter sets in, porcupines converge on places like The Crevice—places that are often easy to spot on a topo map. Here it's the evergreen leaves and twigs of hemlock that keep porcupines fed through winter. In the summer any porcupines up at The Crevice are more likely to be digging for underground false truffles than eating hemlock needles. But in winter they climb high up in the trees, as far out on a branch as is safe, and feed on hemlock. With their sharp front teeth—four rodent incisors—they cut through the twigs. They munch on some leaves, then let the twigs fall.

Porcupines are choosy about which hemlocks they feed on—individual trees within a species vary in their taste due to genetics, site conditions, and defensive compounds. When you come across the one-foot-wide rut in the snow marked by the blunt, pebbly textured feet of a plodding animal and dribbled with piney-smelling green pee, follow it. It will lead you through the forest, past hemlock after hemlock, until you reach the one hemlock, the tastiest of all, where the porcupine feeds day after day, year after year. Perhaps this one tree is also particularly easy to climb.

Hemlock twigs are littered around the base of the tree, each about the size of Japanese foldable fans, maybe a bit larger, but feathered with tiny, dark green hemlock needles. Pick up one of these branches and you'll see the broken end, a bit thinner than your pinky, showing the characteristic 45-degree angle cut of a large rodent. Looking up in the tree, we see denuded branches with strange tufts of green at the ends but no green along the lengths. The teetering porcupine, afraid of venturing out too far on each branch, prunes it like pom-poms on a poodle.

INVADERS

I'm worried about the porcupines at The Crevice. Not because of shovel-wielding landlords but because the woolly adelgid is killing

Figure 8.4.
Hemlock woolly
adelgid.

off our hemlocks up and down the eastern United States (fig. 8.4). For millennia the aphid-like insect has feasted on hemlocks in the western United States, Japan, and China. Those hemlocks evolved ways of defending against the insect, and the trees don't seem to even notice when they're covered with the white fluff. However, the two species of hemlocks in the eastern United States didn't evolve alongside woolly adelgid. So when people accidently transported some woolly adelgid from Japan to Virginia in the mid-twentieth century, the eastern hemlocks' nightmare began.

Over the course of eighty years, woolly adelgid spread to Georgia, Maine, and beyond. Hemlocks down south have been particularly hard hit—like the forests of snags we encountered when searching for chestnut. Luckily for the Northeast, the introduced strain of woolly adelgid came from southern Japan, where the climate is relatively warmer. The invading adelgids aren't adapted to the cold. So up north, when we get a hard winter, it kills off the

Figure 8.5. Elongate hemlock scale.

adelgids, holding them at bay. But as the planet warms and our winters become less severe, the adelgid will inevitably march farther northward.

The good news for hemlock is that, unlike with American chestnut, hemlock isn't that valuable for timber. Which means that timber companies haven't rushed to cut down all the hemlocks before they die—as they did with chestnuts. Hopefully, there will be a handful of hemlocks that, due to just the right set of genes, survive the first wave of adelgids. The next generation of hemlocks would then inherit these genes as they start to reclaim the forests.

Over the past year I've hiked up this mountain many times to photograph adelgid for the chapter's opening image. But I've failed every time to find a single adelgid—is this good news? Instead of adelgid, all I find are heaps of another tiny invasive pest: elongate hemlock scale (fig. 8.5). I've been told that the scale doesn't kill the hemlocks as directly as the adelgid, but there is a complicated interaction with the insects competing against each other and both draining the tree's resources. At a recent Woodsy gathering, I was venting my frustration at failing to capture an adelgid photograph. Then Jesse reminded me of how the last couple winters the polar vortex has slipped down into New England causing weeks of negative temperatures. Those cold snaps must have driven the adelgid back, however temporarily.

The porcupines at The Crevice are rooting for the hemlocks to survive. Having tracked many a porcupine around here, I came to the belief that porcupines depend almost entirely on hemlock in the winter. But the truth is that porcupines can eat pine needles, sugar maple twigs, basswood buds, beech bark, oak bark, and more. Out in New Mexico, Charley and I found a porcupine crossing through the sandy desert at White Sands National Monument, far from any hemlocks.

Biologist Uldis Roze has spent years following porcupines around, naming individuals and getting to know them. Porcupines, he found, all have individual tastes. Two porcupines living side by side might eat very different things. The porcupine Uldis named Rebecca, for instance, loved basswood, whereas Finder loved beech, Squirrel loved sugar maple, and Moth loved hemlock. Why? Well, one reason might be that these trees all have toxic defensive compounds, and the microbes living in the porcupines' guts have to be specially trained to break down these compounds. If Rebecca suddenly started eating hemlock, her gut bacteria might revolt, and she'd be unable to process the tannins that hemlock produces.

So although I think of porcupines as only eating hemlock in the winter, that's only true of the porcupines I know—the ones who tend to hang out in hemlock-rich areas in winter. Even if we lose all our hemlocks, porcupines as a species will survive, but perhaps not the ones I know.

EXPERIMENTAL FORESTS

This summer I followed Sydne and three students in to one of the experiments at Harvard Forest. We walked through a young stand of hemlocks casting a dark shadow into the open forest understory. Then we encountered a sudden shift. Now we were clawing our way through a bright green sea of black birch saplings. Above, the sky was full of dead hemlocks. At the base of each tree, the trunks were encircled by the grooves of a chainsaw.

Sydne and the students fanned out to collect pitfall traps—little plastic vials sunk into the soil and filled with soapy water—to study the ants that fell in. This forest is a simulation of what would happen if woolly adelgid killed off all our hemlocks. Instead of millions of tiny fluffballs sucking at the needles, researchers had girdled the trunks to cut off the flow of sap (fig. 8.6).

Figure 8.6. Hemlock removal experiment: (a) girdled hemlocks surrounded by black birch, and (b) student Nia Riggins sampling ants in the understory.

Hemlock is a typical old-growth species—slow growing, long lived, and shade tolerant. Young hemlock saplings will hang out for decades in a forest understory, slowly growing to take their turn in the canopy. In the dark shadow of a hemlock forest, mid-successional species such as oaks can't survive.

There is an early successional species, black birch, which is also often associated with old-growth hemlock stands. In a stand of hemlocks, one of the ancient trees crashes down. A tiny seed of black birch finds its way onto the fresh dirt around the upturned hemlock roots. The young tree quickly shoots up into the gap, lives for maybe a hundred years, and dies. Meanwhile another hemlock is slowly working its way up for its own 500-year life in the forest.

With the hemlocks gone from the system, everything shifts. Sunlight pours in. Cool streams become warm streams. Moist soil becomes dry soil. Even the ants, as Sydne has found, are profoundly different.

After Sydne and her students finished collecting the first twenty-five ant traps, we walked through a fenced-in area to collect more traps. The fence was built to keep moose and deer out.

Deer are another species, like porcupines, that gather together in hemlock forests in the winter. The dense needles of hemlocks make great shelter, and snowpack is thinner under hemlocks. More often than not, when I find a winter deer bed, shaped like the imprint of a two-foot-long lima bean melted into the snow, it's underneath a hemlock. The same is true of moose beds, though the lima bean is even bigger, as is the sheltering tree. After a good sleep under the hemlocks, deer will wander in search of food. If there's not too much snow on the ground, and if there are some good oaks around, the deer might be able to scrape up acorns from the forest floor. But when the snow is too deep or icy for hooves to reach the acorns below, deer, like porcupines, will feast on hemlock needles and bark.

In some places, where you exclude deer, the forest is completely different. In graduate school I had a little project studying how sal-amanders on islands in lakes are different from salamanders back on the mainland. On the mainland, I'd walk through open forests with scattered hobblebush shrubs. When I'd get in my boat and float over to the islands, I'd encounter impenetrable thickets of American yew. Yew is a favorite of deer, who presumably devoured it all on the mainland but never visited those islands.

What's the role of deer in a forest? In addition to eating vegeta-tion, they eat acorns. Deer presence thus reduces populations of mice, who also rely on acorns. Mice, more so than deer, are the key vector for Lyme disease, so fewer mice means less Lyme dis-ease. Mice also eat spongy moth larvae. So when mouse popula-tions crash, spongy moths explode and eat up all the oaks. This means fewer acorns, and so fewer deer. It makes my head spin, and we haven't even brought in the predators like foxes and coy-otes yet. These are the kinds of complex interactions that research-ers with fences around the country have been working out for de-cades. However, at Harvard Forest it turns out that the populations of deer are so low that the forest inside the deer fences is basically identical to the forest outside the fences.

These big experiments are just a few of the many strange things in the woods at Harvard Forest. Harvard Forest is part of a network of Long Term Ecological Research sites around the coun-try focused on doing research that requires a long investment in a place. For instance, up at a sister site in New Hampshire—Hub-bard Brook—researchers have been studying the various parts of a single watershed from all different angles for decades. Recently, Hubbard Brook researchers brought out firehoses in the dead of winter and simulated an ice storm. The trees, coated with glisten-ing ice, began to snap. What snapped when and how will the trees recover? The answers to these questions will be used by weather forecasters to predict ice storms and by ecologists to predict how forests will respond to climate change.

Figure 8.7. Trees felled by windstorms: (a) pines knocked down in the 1938 hurricane in New England, lying parallel to each other, and (b) blowdown from a 2006 tornado, photographed in 2009, showing varied log orientation. (b: MassGIS, Fugro Earthdata, Inc.)

Elsewhere on the Harvard Forest property is a hurricane simulation. Audrey Barker Plotkin, a forester and ecologist, recently took me and two students there to sample falling leaf litter. This was the day before the town where I live—home to the mountain of conglomerate that holds The Crevice, the porcupines, the oak forest, and the hemlock forest—was to hold a big parade celebrating the town's 300th anniversary. On the drive out to the hurricane simulation plots, Audrey told us about how, eight years ago, they celebrated the twentieth anniversary of the hurricane experiment. To which I asked, "Did you have a parade?"

Twenty-eight years ago Harvard Forest researchers drove out into the forest armed with a huge steel cable and a powerful winch. In one two-acre area, they toppled 276 trees, all toward the northwest—which, according to the researchers' calculations, is the di-

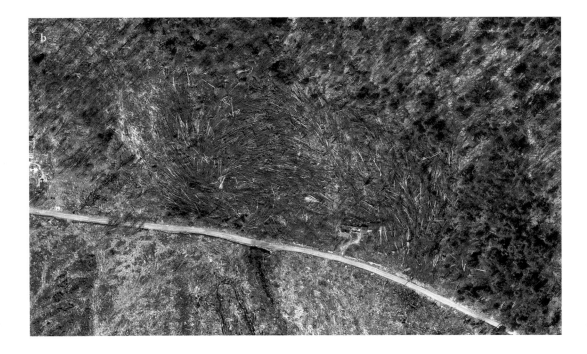

rection of the strongest hurricane-force winds in this region. Although humans always seem surprised and devastated when
hurricanes hit, the forests are used to them. At Harvard Forest, researchers figure that each tree in the forest should expect to be hit
with some sort of damaging hurricane winds every decade or so.
Every hundred years or so, the winds are going to be so bad that
much of the forest will be flattened—as happened in the "Great New
England Hurricane" of 1938 (fig. 8.7).

In the control plot, where researchers left the trees untouched,
Audrey led us into a forest of red oaks. They were decent-sized, but
not huge, and all pretty similar to each other. It struck me as a sort
of young, boring forest—with a predictable history typical for the
region. In the 1800s it was a farm. It was abandoned. Pines grew in.
An oak understory began. The pines were cut. The oaks took over.

Figure 8.8. Hurricane
experiment at Harvard
Forest: (a) hurricane
simulation plot, and
(b) control plot.

But over in the hurricane plot, it was more interesting (fig. 8.8).
In some ways the forest looked much older—almost like an artificially created old-growth forest. There was structural complexity—something more typical of older forests. In undisturbed old-
growth forests, the trees don't all die, or start growing, at the same
time. Individual trees fall over while their neighbors stand tall,
creating a little gap in the canopy. This creates a tapestry of new
gaps, older gaps that have been filled in, and old trees still holding
their ground. And on the floor of an old-growth forest you'll find
big logs in all stages of decay.

In the hurricane simulation there are logs lying all over the
ground. Instead of the dominant trees all being the same size, like

in the control plot, there are two size classes in the canopy. There are, of course, all the young trees that have grown up in the last twenty-eight years since the simulated hurricane touched down. But then there are scattered trees that lived through the hurricane. Relative to the rest of the forest, these trees are huge. Even compared to the trees in the control plot—which are the same age—the trees that survived the hurricane are bigger. That's because, twenty-eight years ago, all their competition was wiped away. Left alone to soak up all the sunlight and soil resources, these trees grew like mad.

It's neat to see a forest with old-growth characteristics around here, as most of our forests are still recovering from the massive land clearing over the past few centuries. And it's not just the trees—the animals, too, were cleared from this region due to the lack of forests, intense game hunting, and high bounties on predators. We lost heath hens, passenger pigeons, and sea mink—gone forever. We lost all our wolves and cougars—although they survived elsewhere. We lost almost all our bear and deer, and they have since rebounded from holdout populations, with the help of active management aimed at restoring their populations. We lost all our beavers—but then the state reintroduced them further west in the 1930s and 1940s. By the 1970s, beavers had spread up and down the Connecticut River Valley, and by the late 1980s they had repopulated most of the state.

Standing out in the forest with Audrey and the students, I inspected some of the red oak logs that were pulled down twenty-eight years ago. How do they compare to the logs in the bright forest up on top of our hill where I took the *Field Naturalists*? I'd brought my laptop out to the hurricane plot so that I could look up all the Harvard Forest data on each numbered log we encountered. I pulled up a photo of the logs from the bright forest to compare the amount of decay in those logs to the ones in the hurricane plot. It was hard to say, but they could have been somewhere around the same age.

Well, maybe the logs in the bright forest are a bit older. To explain my reasoning, I told Audrey that the forest in my book chapter is an oak-hickory-hop hornbeam forest. "Yup," she said knowingly. These types of forests, which grow on south-facing slopes, are characterized by hot, dry conditions. In contrast, the moist conditions of the forest in which we were now standing make a much better place for moss and critters to grow on and in the logs, which therefore deteriorate much faster.

Audrey pointed out the root ball of one of the upturned trees in her plot. When the tree was tipped over, the living roots held fast to the soil and rocks, transporting a huge pile of dirt out of the ground. Where the tree once stood was now a big pit. To the side, where the exposed roots came to rest on the ground, now sat a big mound of dirt (fig. 8.9).

Then Audrey pointed out another log. This one, however, had no pit or mound at its end. The roots at the base of the tree all seemed to have broken off close to the trunk. This was not a tree that the researchers pulled down. Audrey had a theory. If trees are weakened or already dead when they fall over, their roots break off. Sometimes the trunk snaps above the base, and sometimes the break occurs just underground. Without roots clinging to dirt, such falling trees don't make pits and mounds. It's when the lives of healthy trees are cut short by a violent storm that the roots are strong enough to create a pit and mound.

So what role do these pits and mounds play in the forests? Armed with a tape measure and clipboard, Audrey has been crawling for years through the old hurricane experiment measuring the dirt piles. "Charismatic micro-topography," we joke. Over the last two decades, Audrey and other researchers followed the erosion of individual pits and mounds as the mounds slowly grew lower and wider and the pits slowly filled with leaves and soil.

Pits and mounds are a common feature of old forests. Depending on the local climate, they can last for a few decades or a few thousand years. Collectively, pits and mounds on the forest floor record generations of trees growing, falling, and rotting into the earth. In temperate forests, if you don't see pits and mounds, you have to wonder, why not? Often it's because the land was once cleared and plowed flat by people. The floor beneath virgin old-growth forests is typically an irregularly undulating surface of pits and mounds in various stages of decay.

Out in the hurricane experiment, Audrey watched over the years where little seeds landed in the dirt on top of the mounds, sprouted, and grew up into small saplings. For most trees, especially black birch, those mounds are the best place in the forest to

Figure 8.9. (a) Pit with breeding wood frogs next to a freshly tipped-up root ball of a tree in Pennsylvania: wood frog egg masses appear as dark splotches emerging from the water, near the fresh dirt. (b) An old pit and mound in Tennessee.

grow. It's prime real estate high above the competition, with exposed soil that isn't smothered by leaves, and a rotting tree below to provide nutrients to the growing sapling. Plus black birch is an early successional species that needs light to grow, like the light that pours in through the gap in the canopy where the old tree once stood. These little gaps where trees fall are why old-growth forests aren't simple monocultures of late-successional species.

After many decades the old, toppled tree and its root ball rot away to almost nothing. But the aging black birch still keeps a record of the mound it sprouted on. The birch stands on a set of funny spread legs that emerge from different points in the ground and fuse together a few feet up to form the main trunk. Enclosed by these legs is an empty space—the shape of the old root ball. In some forests,

like those of the Pacific Northwest, the whole trunks of fallen trees become prime real estate for sprouting trees, such that trees often grow in lines marking the location of old "nurse logs" (fig. 8.10).

Audrey concluded that the mounds themselves promote diversity in the forest. Without the mounds, species like black birch wouldn't have a place to grow in the forest. The pits, however, seem less exciting to Audrey. In the Harvard Forest hurricane simulation, most young trees didn't like to grow down in the pits.

But, as a herpetologist, I love the pits. I root around through the layers of leaves trapped in the pits hoping to find an overwintering box turtle or a hiding ring-neck snake. In lowlands in early spring, I leap from pit to pit, hoping to find some filled with water. Hearing a cackling sound echoing through the forest, I home in on

Figure 8.10. Nurse log in the Smokies.

one water-filled pit overflowing with breeding wood frogs. Without these pits where would all the reptiles and amphibians go?

DISTURBANCE

At last the *Field Naturalist* students have arrived at the answers to the little mysteries.

The circle of moss? That's the mound from the root ball of a tree that toppled long, long ago and has completely returned to soil.

That dead sapling? You can see exposed growth rings where for several years the bark tried to heal over a wound, creeping sideways across the stem. In fact, that thick scar is the only bark that's still attached to the tree—the rest of the bark below and above has sloughed off. Hemlocks do that. A few other species do also, but the feathery twigs tell us it's hemlock. So what injured it? A porcupine? Porcupines do like hemlocks, at least in winter. But a little sapling like this—exposed on this hill—doesn't scream porcupine. It's such a skinny trunk, and there's no evidence of angled cuts on the branches. Moreover, there's only a small portion of the stem that actually shows evidence of scarring.

No, this tree doesn't seem like it died because something ate it. The lone hemlock sapling in a hardwood understory. At the top of a hill, in a forest littered with acorns, and a short walk from the cover of much denser hemlock. This is a perfect place for deer to hang out in winter, and the sapling is a perfect signpost on which to make a territorial claim. This is a classic, albeit very old and weathered, marking sign of a male deer thrashing his antlers against the sapling.

What do we see in those understory hemlocks? For one thing, there's a sharp horizontal line. Above the line, a sea of green hemlock needles. Below the line, a dark void. This is the browse line of a voracious herd of deer, out of nuts, munching every needle within reach. The other pattern worth noting is that, while the understory is all hemlock over there, the canopy, as told by the leaf litter, is all hardwoods. If left alone, succession would turn that little hardwood stand into hemlocks.

The branches on the ground? Looking around, branches of that size, shape, and degree of decay are littered about this forest. They came from the tree tops and were all brought down at once. In an ice storm—a particularly bad one ten years ago.

The rotting log? That's where a tree fell. What else is there to say? Well, a close look at the wood shows that it's a red oak for starters. Also, there are several other logs of the same size, with the same amount of decay, all lying in the same direction. Three root balls are visible in the top left photo of the chapter's opening image, and, like the shadows of the setting sun, the logs are all pointing eastward. That suggests that a great gust of wind toppled the trees at once. A hurricane? A tornado? A winter nor'easter? A summer thunderstorm?

A tornado would scatter trees willy-nilly. The rotational forces of a hurricane moving north up through New England typically topple trees to the west around here. A winter storm moving in from the Northeast would also knock trees down toward the west. But thunderstorms, which tend to march from west to east, tend to knock trees over toward the east. So, chances are, a strong gust, known as a microburst, from a severe thunderstorm knocked these red oaks down. I'd guess that these trunks have been lying here maybe thirty or fifty years, so that storm would have happened in the late twentieth century.

Looking at the roots on these overturned trees, they don't seem to have made a mound around them when they died. No, they seem to have broken off near the base when they fell. Were they already dead when that storm hit? Well, it was nearly forty years ago, in

1981, that the spongy moth outbreak killed off Bill Patterson's oaks on the other side of the slope. I bet they killed these oaks too.

But there's more that these oaks have to tell. The trunks are all forked. That suggests that these were the regrown stems of trees that had been previously cut down. Looking at the size of the forked trees, I'd guess they'd been growing for maybe a hundred years or more, which would put the logging event in the mid-1800s.

How big were the original trees that were cut down? If we estimate the centers of the base of each trunk on the tree, the distance between them is about the original diameter of the earlier trunk that was cut down—since the new sprouts start growing from the edges of the cut stump. In this case the original trees were about two feet in diameter. So these original oaks probably started as acorns in the mid-1700s.

HEADING HOME

As Juno and I stare at the hemlock slice in our living room, ticking off the decades, our count takes us past 50 rings, past 100 rings, then past 200 rings. The ring at the center of this slice of hemlock formed in the year 1776, around the time that the red oak on the hill above was a young sapling. The oak was felled twice since then, while the hemlock just kept growing older. And, as I remember, this slice of the hemlock wasn't cut from the base of the tree but from fifteen feet up along the trunk. It likely took the hemlock several decades to reach that high, making the tree well over 250 years old. Older, perhaps, than all the old towns around here. It's probably time for this forest to have a parade in its honor.

So why is this old hemlock here? There are two answers to this question. First, this north-facing slope makes for prime hemlock habitat. Although these trees can live in a variety of settings, this is one that they're especially good at. The oak forest above, especially the part with no understory hemlocks, faces south, catching the hot, dry sunlight all day. That hill of oaks casts its shadow onto this hemlock ravine. Thus, it is dark and relatively humid here—great for hemlocks.

Up on the hilltop the trees are short. In part it's because the dryness makes for difficult growing conditions. But it's also because the trees up there take a beating from the elements. When storms hit, ridges often take the brunt of the wind—although not always. And from freezing to baking, the temperatures are more extreme up there. The protective snowpack melts faster on the sunny ridges, leaving roots in the soils exposed to damaging cold at night.

The hemlocks in the ravine, however, are sheltered. Beyond just the weather extremes, the steep slopes have also protected this forest from humans. On such steep slopes, logging is difficult. Farming is even harder. Plus, although hemlock was historically prized for tanning hides, if you try to make a board out of it, the wood tends to fall apart. So loggers in this area would have left hemlock alone.

This is the truth of most remaining old-growth forests. They are in the hard-to-reach places. That's why, when we were looking for chestnuts in North Carolina old growth, we had to travel up the Tail of the Dragon—that ludicrously sinuous road—to get there. And that's why I put in the extra effort to push the children up this mountain. In the Northeast, old growth is often concentrated on steep, north-facing slopes of hemlock. Having survived the centuries primarily because people were uninterested in these forests, the arrival of woolly adelgid is suddenly threatening the ancient hemlocks. Indeed, the hemlocks here at this site are falling apart, and it's not clear how much longer they will last.

The sun is gone from this valley now. Back in the stroller, Juno and Alder huddle together for warmth like hibernating snakes as we roll down the mountain.

MAJOR LESSONS FOR INTERPRETING A LANDSCAPE

- What are the disturbance dynamics in your site? Think about windstorms, herbivores, pathogens, and so on.
- Where are the local microhabitats that support wildlife through winter or the dry season?
- How does topography drive patterns in your landscape at large and small scales?

(overleaf)
Figure 9.1. A clearing in the forest.

9. Relics

I'm staring at a footprint. It's big and obvious. There's a long, alternating string of them that continues for perhaps twenty yards or more. The track size and stride length aren't too different from my own. But my feet don't make the slightest impression in this firm litter of hemlock twigs and needles. I'd have to really twist and dig in with my heels. Is this what I think it is?

I wander off, then return. I search for a bit. Then right there on the trail I find a big tree that's been gouged all over. An hour before, on my hike in, I had breezed right by it, along with all the mangled saplings that surround the tree. It's what trackers call a "whammy tree." The string of footprints leads right to it.

This is the ritual trail of a bear. "Ritual" because the bears methodically slip their feet into the same footprints each time they visit the tree, twisting with each stomp to rub in their scent. As I tell my kids, it's where the bears come to dance.

What a perfect spot for this sign: in the center of a low, long, forested hill that juts like a peninsula into a surrounding habitat. When I was sleeping a few feet away in my old debris hut—now nothing but dirt—I wonder, was this sign here the whole time?

It's now time to head for that other habitat.

Wandering among six-foot boulders beneath a dark hemlock canopy, I catch sight of light pouring into the forest up ahead. I cut down a short slope, scratch my way through a wall of dense

Figure 9.2. View of dead snags on south end of clearing.

shrubs, and am confronted with an expansive view (fig. 9.1). What is this place, and why is it so starkly different from the surrounding forest? After soaking up the warm sun on this chilly December afternoon, my gaze turns to the forty-foot dead trees standing about. On the far side of this clearing, there is an entire ghost forest of these snags (figs. 9.2 and 9.3). Why are there so many tall dead trees but none alive?

What is a clearing? Having lived most of my life in the eastern forest, I expect to see trees wherever I go. When I find myself in a place without trees, I naturally wonder what is holding them back. Is it from logging? Cattle? Fires? Floods? Hurricanes? If we venture across North America west to the Great Plains, desert, or mountain biomes, trees are naturally prevented from growing by combinations of lack of water, grazing mammals, fire, and other forces.

Perhaps if you come from Nebraska, the sight of forest might cause you to wonder what is holding back the prairie—a fair question. Try planting a prairie here in the East and leaving it alone. As we saw at Maggie's Forest, succession will eventually bring in an army of trees that block the sun and wreck your prairie. To keep your prairie, you must regularly disturb the landscape by unleashing cattle, mowers, or fire into the field to kill off the tree saplings.

Trying to maintain an open field in the eastern forest can be like trying to maintain a yard of big water-loving trees in the Desert Southwest. However, that's not to say that such trees never occur naturally in the deserts—if you go down by a river, you will find stands of cottonwoods. The same is true in the eastern forests—if you know where to look, you will find natural meadows.

Figure 9.3. (a) Aerial photo of the chapter site, and (b) topographic map of the chapter site. (a: MassGIS, Sanborn LLC; b: USGS)

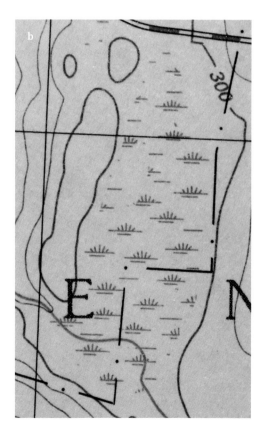

This site reveals a complex pattern of vegetation types and heights spread about in organic clumps. I don't see uniform vegetation, neat lines, logging stumps, fences, or heavily worn trails. This suggests that we are looking at a natural clearing in the forest, not one of human creation.

CATTAILS

Do you see the cattail peeking out of our chapter-opening image? The sight of cattails transports me to their signature habitats: nutrient-rich marshes inhabiting backwaters of streams, on the shores of lakes, and in artificial retention ponds. I close my eyes and see myself on a sunny winter day clambering through a nearly impenetrable cattail stand. As I step carefully forward in my imagination, I hear the crackling sounds of brittle stalks breaking while snow crunches between my boots and the ice-covered water. Without warning the ice breaks, jamming against my shins as I drop through to the mud below. My boots fill up with water. When I stand still, the water warms in my socks. But with every step my boots make a sucking sound, and I feel cold on the soles of my feet from new water rushing in.

I plod along the meandering trail of swimming muskrats. Muskrats love the taste of cattail just as I do. The spring stalks taste exactly like cucumber. But it's the winter seed heads that to me are the most magical part. My students begin giggling with delight as they pluck at the firm brown corndogs. Out of this tight bundle, golden velvet masses of a quarter million seeds explode into the wind. It's that texture I love. Each seed is fashioned like a helicopter, with a payload suspended beneath a whirl of blades that unfold as the seed crawls out of the mass. The seeds tumble out like so many clowns packed in a car. How did they all fit in there? The furry floating clowns now completely fill the air. They coat my nose and tickle it, causing my head to recoil with a shiver as I open my eyes and return to the present.

This image before me bears little resemblance to the landscape I saw when I closed my eyes.

Yes, this cattail seems to be telling us that water is to blame for the lack of big trees here. To survive with roots under water year round, trees need special adaptations, such as the snorkel-like "knees" of mangroves that stick up above the waterline to help the roots breathe. Such trees are not abundant in this region, so when soils are deeply waterlogged all the time, there are usually no large trees—as in many lakeside marshes.

Indeed, ask the small white pine sapling on the right side of this chapter's opening image why it is a sickly yellow color. This pine may say that it found purchase in dry soil for a few years next to the big rock, but now it's too big to grow sufficient roots without drowning. But weren't there big trees and woody shrubs in the Great Swamp? In that case the water was held at the surface by a layer of clay, "perched" above the natural groundwater level. The water layer was thin, seasonal, and, during the wet season, tree roots could find air in little mounds above the water layer or conceivably even in air pockets beneath the clay layer. The lack of trees where I am standing now may mean that this wetland is deeper and more permanent than our tupelo swamp.

But this is not the dense cattail marsh I waded through in my memory—there are only a few scattered emissaries of this species. What other friends can we find here to offer more clues?

PITCHER PLANTS

To my eye the most striking species is the pitcher plant in the foreground of our chapter-opening image—seen as a rosette of red leaves beneath dried flower stalks. Here we meet one of those incredible carnivorous species that few realize live among us. The "pitchers" are hefty leaves that have been modified to hold water.

I imagine shrinking down to fly-size and perching on the pitcher lip. Peering down over the edge, careful not to fall, I can see insects floating in the dimly lit water near the bottom. Many of these insects were drawn in by the sweet nectar on the rim of the pitcher. Looking to my left, I see an ant, about my size, drinking that sugary liquid. That nectar looks so yummy and barely within reach. I stretch my arm out to grab just a little taste of the delicious nectar. Teetering, I make a quick adjustment to my balance and bend down to steady myself on the lip of the pitcher. But my feet slip, and I find myself scrambling to hold on to the downward-pointing hairs that line the walls of the pitcher. After a moment of panic in mid-air, I splash into the water.

I come up gasping and begin to tread water among the other floating bodies. Slowly, these prey are being consumed by the digestive juices secreted by the plant. Some are consumed by the other organisms who call the pitchers home. Enclosed by maroon walls, here is an entire gloomy food web of bacteria, protists, nematodes, fly larvae, ants, spiders, and their acquaintances. I'm not the first vertebrate to fall in here; researchers out of Harvard Forest recently reported that this species of pitcher plant consumes newts on a somewhat regular basis. In fact, the last time I took my students to this wetland, they tore open a pitcher, just one, and shrieked in horror and delight when a rotting redback salamander corpse came rolling out (fig. 9.4). That could be me.

To a pitcher plant the critters it consumes serve as vitamin pills. When your doctor says you're not eating enough calcium, she may advise you to take some calcium supplements. Ancestral pitcher plants found themselves in nitrogen-poor environments, and so Dr. Evolution prescribed nitrogen-rich insects, to be taken only when needed. If we sprinkle a little nitrogen fertilizer on the

Figure 9.4.
Salamander eaten
by pitcher plant,
found by *Field Natu-
ralist* students.

roots of this plant before us, she will stop growing pitchers. No
more nitrogen deficiency? No more need to swallow insect pills.

Why would any wetland soil lack nutrients? When a wetland
receives most of its water from rainfall, it doesn't enjoy the nutri-
tional benefit of groundwater percolated through rich soils. This
results in an acidic, nutrient-poor wetland perfectly suited for
sphagnum moss (fig. 9.5). Sphagnum, in turn, enhances the acidic
conditions of the water, with its complex minute structures acting as cation exchange surfaces and releasing more hydrogen ions

Figure 9.5. Close-up of sphagnum moss.

into the water. Thus, an acidic bog is formed, making a home for pitcher plants and other leafy carnivores (fig. 9.6).

A bog is a type of peatland—a wetland that develops layers of organic peat because the rate at which plants grow is faster than the rate at which plants decompose. Peatlands are thus important global carbon sinks. The peat itself is primarily composed of sphagnum moss, with new life growing on top of old, dying layers. Long harvested and burned as a source of fuel, peat has been promoted as a renewable energy source by advocates in peat-rich countries like Finland. However, peat grows very slowly, at about $\frac{1}{10}$ of an inch or so per year, hardly fast enough to be considered a sustainable source of energy.

There are many different types of peatlands, occurring in varied places and contexts—from tropical forests in Malaysia to Venus-flytrap-harboring wetlands in the Carolinas to the boreal string bogs of Labrador. Their formation can unfold in multiple ways. Sometimes they begin as isolated basins, such as kettle ponds—

which are holes left in the landscape by giant chunks of melting ice dropped by retreating glaciers (see pond in fig. 3.2). Sphagnum may begin to grow across the pond surface and slowly transform open water to a mat of wet vegetation. Sometimes bogs form from parts of wetlands that become hydrologically isolated from the rest of the wetland. Sometimes a sphagnum wetland may form by piling up layers and creeping out across dry land. In some places acid-forming bedrock or mine tailings will trigger formation of bog-like communities.

Standing in this spot and then looking down, you might be surprised to find that, where you thought you were standing on solid ground, your feet have sunk a few inches into a squishy wet mat of sphagnum. Holding a long stick, the *Field Naturalist* students probe

Figure 9.6. Thread-leaf sundew eating a bluet in New Jersey.

the surface out beneath that mysterious forest of dead snags. The stick easily slides down into the sphagnum five feet or more before stopping.

Once when I was surveying a small pond for salamanders, I stepped onto what seemed like solid ground only to find myself suddenly waist deep in layers of sphagnum. My first attempts to wriggle out caused me to sink even deeper. After a few panicked minutes, I managed to reach a nearby branch and drag myself back out to solid land. With less luck, perhaps I would have become another of the many "bog bodies," buried in peat and discovered thousands of years later, preserved by the extreme acidic conditions that inhibit decomposition within the wetland.

Slowly growing peat not only preserves bodies and stores energy but also maintains a historical record of the conditions present when the layers formed. Digging deeper and deeper corresponds to looking farther and farther back in time. This is how paleoecologists, inspecting pollen grains stuck in the layers of peat, can reconstruct the past—like Margaret Davis mapping the movements of tree species over the past 20,000 years.

In January, on a −6°F day, David Foster took me for a walk out to a couple of swamps at Harvard Forest. David, the esteemed director of Harvard Forest, is an expert in reconstructing ecological histories, and I wanted to learn about his work coring bogs. On the walk out I figured I'd try to impress him with some of my own expertise, so I dropped to my knees to sniff some yellowish snow.

"Gray fox," I said to David.

He looked back, stone-faced. I persisted.

"You know the smell of red fox?" I asked, eagerly filling in the silence. "Red Fox is skunky; you'd smell it from far away. Gray fox is milder, more like fisher, if you know fisher pee."

Unimpressed, David told me that he hadn't had much experience in smelling pee. So I shifted gears and pointed out some fisher tracks crossing the road.

David walked me through a spot where dozens of dead chestnuts were leaning up against hemlocks—as they had been since 1913. Just before the blight hit, a researcher had carefully marked and measured all the chestnuts in this forest to understand what characteristics would allow a tree to survive the blight—assuming some would survive while others wouldn't. Then the blight hit, and every single chestnut died.

At last we reached the wetland called "Hemlock Hollow," which looked to me like a small vernal pool. Decades ago, when David first arrived at Harvard Forest, one of his students made bathymetric measurements of the pond and discovered a deep bed of sediments below. So they took some sediment cores. Back in the lab, the samples showed a layered sequence from which they were able to date back an 8,000-year chronology. In a neighboring wetland David pulled another core out that went back nearly 12,000 years.

The sediments in these layers showed how the ponds had evolved. The deepest layers in the basin were sand, from the recently departed glaciers. Then bluish-gray clay and green pond bottom algae reflected calm, open waters. Then peat showed when each pond had filled in to become forested wetlands. Digging many more cores, Rebecca Anderson later continued this work and showed that after the ponds had turned into forested wetlands, sphagnum crept out over the edges as the wetland expanded laterally uphill.

Wedged into these sediments was ancient pollen which had landed on the water's surface before sinking to the bottom, and which now tells the story of the broader landscape. Looking through the microscope, David picked out the species of pollen in each layer, matching it with radiocarbon dates. While hemlock has been dominant in the forests surrounding these wetlands for about 8,000 years, it was only in the last couple thousand years that chestnut has been around. The deepest layers showed an open land of sedges—tundra following the retreat of the glaciers.

Just above these, about 10,000 years ago, the landscape was dominated by spruce—like the forests we now find north of us today.

Spruce, like the little lone tree now standing at the left of our chapter-opening image, just beyond the pitcher plant, not far from the cattails. I wonder, is that a black spruce or a red spruce? Here, south of the northern boreal forest, black spruce occurs primarily in acidic bogs, whereas red spruce occurs in more neutral wetlands.

Whenever I see a black spruce in this state, I can hear Charley mumbling in my head about dwarf mistletoe—we better go over and take look to see if we can find it. This tiny parasitic plant is related to common mistletoe but stands no more than an inch tall. If we find dwarf mistletoe, we can report the occurrence of a species on the state's endangered species list. Its rarity is largely because the primary local host, black spruce, occurs so infrequently here—farther north, dwarf mistletoe is considered a pest species that managers target for eradication. But in this state, dwarf mistletoe is known from only twenty current locations, although perhaps the low number of records relates to the fact that it is easily overlooked, and too few people stomp out into bogs to inspect black spruce branches.

Over at the spruce, the dark, shiny needles, the reddish twigs, and the lack of old cones suggest this is a red spruce. No mistletoe here.

So what should we call this place? A bog? We see sphagnum moss and pitcher plants. If we come back in the summer, we will also find carnivorous sundews, which have withered by December. That's not to mention the tawny cottongrass visible in the chapter-opening image, which, like most other cottongrasses, is usually another indicator of acidic or nutrient-poor wetlands. But instead of black spruce, we've got red spruce. And don't forget the cattail. Remember how we said cattails are typical marsh residents and that marshes are defined as nutrient-rich wetlands? Also have a look at the dense stand of *Phragmites*, or common reed—the invasive type here. This is not a typical bog species. Finally, look at the aerial photo (see fig. 9.3a)—the wetland is bisected by a distinct stream channel. This does not appear to be a pure, hydrologically isolated bog dominated by bog-only species. Rather, it is more of a patchwork of bog species and other species. This, then, would fall more in the category of a fen. A fen is a type of peatland that has some groundwater inputs and outputs, providing more nutrients than would be available for plants growing in a bog.

CORNUCOPIA IN A FEN

Beyond their sheer beauty, fens are important homes to a rich set of species, with plenty for everybody. Notice the animal trail cutting through the vegetation in the chapter-opening image. The dense vegetation of such wetlands supports dense populations of small animals, which in turn attract many predators like foxes, coyotes, bobcats, weasels, and others. Large herbivores such as deer and moose will come down to feed on the sedges. Perhaps the moose who pooped on the log came down to sleep in the cold, wet sphagnum. At the southern extent of their range here, moose must resort to such tactics to stay cool enough in this relatively warm climate.

On the far side of this fen is a nice crop of wild cranberries, and the red stems of the highbush blueberry promise bushels of sweet blueberries in summer. I spent one summer wandering this very fen, gorging on blueberries until my belly was full, all the while following fresh tracks of bears who were doing the same. Toward the end of that summer, Charley and I spent a couple days camping out, feasting on nothing but cucumber-flavored cattail stalks, bamboo-like reeds, and other wetland plants. When we awoke by our dwindled campfire in the morning, Charley turned to me and asked, nonchalantly, "Hey, did you see the bear?" I looked up in

Figure 9.7. Bald-faced hornet collecting wood fibers from an old fence—note the stripe of freshly exposed wood above it.

time to see a berry-filled mass of poop falling out, about twenty feet from us, as the large bear casually walked past, never breaking stride. I'm not sure if it was the swampy aftertaste from the previous day's forage, the smoke, or the presence of the bear, but in the same moment Charley rolled over and puked.

With all this life around, even those standing dead trees support a variety of critters. Many people seem to believe that deadwood should be cut down and removed—an outlook that drives me crazy. I view deadwood as the most important and most beautiful habitat that trees provide, depended on by many creatures. Consider cavity-nesting species such as woodpeckers, chickadees, owls, and flying squirrels. Or the bears that tear the wood apart to eat the grubs living inside. If we look closely at the snags in this fen, we will see light marks—just like the little marks on the fence post in Caleb and Maya's yard—where wood fibers were stripped, rolled into balls, and carried by wasps back to their nests (see fig. 1.1). The fibers were then chewed up and mixed with saliva to make wasp paper, which was used to construct the nests that protect the developing eggs and larvae. This is just one of the little creatures making good use of those beautiful snags (fig. 9.7).

But back to the question that started us off: What is the story behind those snags? What is the history of this landscape? If you notice in figure 9.2, the branches on the dead trees are arranged in the classic, evenly spaced whorls that indicate pines—likely white pines. A white pine will grow one whorl of branches for each year that it's alive. So we can count backward and estimate the age of these trees, keeping in mind that many of the whorls have broken off by now. By my count, those trees were maybe around sixty-five years old when they died. Given the density of pines, this seems to have been a stand of mostly white pines, following the pattern of white pines invading an old field after it was abandoned. So what happened to the stand of early successional white pines?

WATER LEVELS

Perhaps you've guessed that rising waters killed the trees. Sure, but why did the waters rise?

Let's take a closer look at how water levels are controlled in a wetland that we suspect is at least partially fed by groundwater. When it rains, some water is taken up by tree roots. If it is a particularly heavy rain, some water may flow rapidly downhill on top of the ground surface. The remainder of the water soaks into the ground and filters down to the water table. The water table is the

Figure 9.8. Simplified version of a water table.

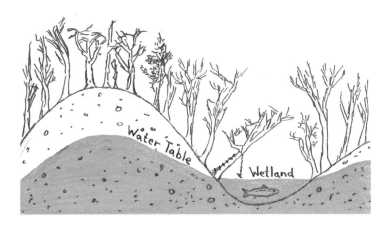

line below which all open pores in between soil or rock are filled with water. If you dig a deep hole, your hole will fill in with standing water to the height of the water table.

As a general approximation, if you look at a profile of the landscape in cross section, the water table follows the overall contours of the land, but with less exaggerated features (fig. 9.8). Beneath a big hill, the water table profile looks like a small hill. As you go further and further downhill, the water table gets closer to the surface. That's a good thing to remember if you're ever in a bind and really need to dig a hole to find water—the water table should be closest to the surface at the lowest point on the landscape. Where the water table intersects the ground surface, you find a river, pond, lake, water-filled hole, or maybe a fen.

If we wanted to raise the water levels of a wetland, one approach would be to increase the amount of water flowing in. Short of causing more rain to fall, if you covered the porous ground with impervious pavement, more surface runoff will flow directly downhill, bypassing the groundwater. This is why streams and wetlands across urban and suburban landscapes become overloaded with runoff and gouged by flash floods—we've removed nature's system for rain absorption.

Another way to increase water flow would be to cut down the trees uphill from the wetland. With fewer trees sucking up groundwater, there will be more groundwater to feed the wetlands. This is why reservoir managers seeking to maximize water in their lakes will often cut down surrounding trees. This concept, I believe, also explains what I found one May day when I went to survey a population of rare marbled salamanders in a small vernal pool. Instead of water and happy larvae, I found an expanse of dry, cracked mud with small, rotting bodies baking in the sun. I think the trees had leafed out too early—before the salamanders were ready to leave the pond. If a warming climate means trees begin to leaf out earlier and earlier, I wonder if we might find more and more populations of salamander larvae baked in the sun.

The other way to raise the water levels in a wetland is, of course, to let less water escape. Once in a wetland, water stays there until it either evaporates, seeps further down through the ground, or flows out through a surface channel. It may be hard to control the evaporation from the surface or groundwater flows, but you could plug up outflowing streams with a dam. Looking at the landscape surrounding this wetland, we see no logging, parking lots, or other evidence for a recent change to the water coming downhill. This leaves us with the explanation that perhaps would have been your initial hypothesis—there must have been some sort of dam put in downstream to raise the water levels and kill the trees.

AGING A WETLAND

Looking at those trees, we might guess that they've been dead for several decades, if that. So perhaps this place was a forest until someone dammed up the stream in the last thirty years or so, right? But wait, how old is this wetland?

Thinking about the number of habitat specialists—pitcher plant, sundew, cranberry, cottongrass—I'd guess that this natural community must have taken quite some time to form. Consider the process of dispersal, using a classic ecological island metaphor. If you construct a new island by dumping a huge pile of dirt in the middle of the ocean, at first there will be no species. You must then wait for species to colonize the island, sending seeds or propagules from other nearby land. If you come back to check on your island, the number of species occupying it will depend on a few factors. In general you will see more species on bigger islands, on islands closer to other landmasses, and on islands that have existed for a long time. This basic line of thought underlies the theories known as "Island Biogeography Theory" and "Metapopulation Theory" that are central guides for conservation planning.

The same ecological thinking applies to terrestrial "habitat islands," such as a park in the middle of an urban metropolis. If you want your urban park to contain high native biodiversity, keep it big and try to protect nearby native species to act as a source for dispersing individuals. With the salamanders in my dissertation, the most important determinant of whether we'd find salamanders breeding at any given pond was not whether there was salamander habitat right around the pond but whether there was salamander habitat a mile away from the pond. That's the scale of metapopulations. If, a mile away, all the forests around the pond had been cut down, it didn't matter whether or not there were awesome forests right around the pond—there would be no source of salamander colonizers to maintain a population at the pond. All that prime salamander habitat would be devoid of salamanders.

Although, connectivity isn't always good. At a recent conservation conference, researcher Molly Bletz stood at the front of the room and described a trip she took to Panama a few years back.

She and fellow researchers were surveying for frogs. As they walked out into the tropical rain forest, they were expecting to be greeted by hundreds of frogs hopping and calling from leaf litter, puddles, trees, and bryophytes. That's what previous researchers had found. But now Molly found only two lonesome frogs chirping worriedly.

At another conference I'd heard a similar story about salamanders in Guatemala. Walking up a pristine forested mountain, untouched by logging or development, researchers flipped logs that in the decades prior had been teeming with salamanders. Now most of the species were completely gone. From an amphibian perspective these had become ghost forests.

Like Native American communities decimated by smallpox, the frogs and salamanders were killed off by a disease from another continent. In this case it was a fungus known as chytrid.

After telling her story Molly went on to describe her current work on US salamanders. The United States is a global hotspot for salamander diversity, and so far they are disease-free—or at least free of this particular disease. But, as we speak, there is a wave of a new chytrid fungus that is sweeping through Europe, wiping out their salamanders. Our salamanders—especially the newts, which are among our most abundant species—are highly susceptible to that fungus.

Chytrid fungus is spread across continents by the pet trade, where many species—including several frogs—commonly carry the salamander disease without suffering from it. With legal battles still raging over whether we can ban trade of these disease-carrying species, all it takes is for one child to grow tired of her fungus-infected pet and dump it in some nearby pond; that could spark the mass die-off of our salamanders.

After her talk I approached Molly to learn more about the looming threat to our salamanders. I asked her, is this a case where connectivity is bad? In response she told me an anecdote about the devastation of the European fire salamanders. The new chytrid fungus was first discovered in a small town in the Netherlands that had lost most of its salamanders to the disease. This was the country's largest and most well-known population of fire salamanders. The town's forest, once crawling with salamanders, was now empty. But, it turned out, there was one previously unknown population a half-mile away in a tiny scrap of forest surrounded by developed lands and farm fields—which function as barriers to salamander movement. And this isolated forest patch was full of healthy salamanders. Likewise, across Europe, forest patches that are surrounded by busy highways are escaping the epidemic. Had tunnels been in place to help salamanders cross beneath these roads, chytrid fungus may have hitched a ride across. In these cases the roads had saved the salamanders. Habitat fragmentation was good.

Still, perhaps because I don't study diseases, I'm mostly of a mindset that fragmentation is usually something to fight against. I think about a small pond in a tiny forest in the middle of a giant cornfield and see isolated salamanders with no way to get help from the outside world. Eventually, something is likely to wipe out the population. If not disease, then perhaps drought, flood, fire, frost, inbreeding, or predators. I imagine they're going to need an occasional colonizer to keep them going.

Metapopulation theory applies to an island in the middle of an ocean, a park in the middle of a metropolis, a forest patch in the middle of a cornfield. How about a fen in the middle of a forest? Suitable pitcher plant habitats are few and far between in the forests around here, and so I imagine we must wait a long time before a pitcher plant seed happens to land in any newly formed fen. How long? Dispersal is one of the most difficult ecological processes to quantify, since it's based on rare events and tiny objects. So it's hard to say how long it took this community to form—but my sense is it didn't happen overnight.

Remember also that we seem to be standing on top of five feet of sediment. If that sediment is all peat, which grows only ¹⁄₁₀ of an inch or so a year, then to build up that much peat should take many hundreds of years. So we'd guess that this fen has been around a long time. But those trees have only been dead a few decades. Perhaps, then, this place has been a wet fen for a very long time, except during a dry spell long enough for a forest to begin to grow.

FOUR-LEGGED SUSPECT

So what may have caused a dry spell in this basin that ended thirty or so years ago? Remember from Chapter 8 our discussion of all the animals that once roamed our landscape before they were killed off by people? There's one mischievous friend whom hunters had eradicated from our area by 1700, but then by the late 1980s had repopulated most of the local landscape thanks to intentional reintroduction: beaver (figs. 9.9 and 9.10).

Before humans were making ponds and lakes left and right, beavers were the primary source of large standing bodies of water throughout much of their range. Ducks, newts, turtles, moose, otters, and a host of others have long depended on beaver ponds. I am often amazed at the volume of water that beavers can hold back with a single dam.

A few years ago I was trailing a moose with a South African animal tracker who had come to New England to trade tracking knowledge. We came on a typical beaver dam, about seven feet high, and he asked, "Who made this?" He thought we were joking when we told him it was made by an animal. It took much convincing, and eventually he just stood there with his mouth open shaking his head. After seeing the beaver's work and later seeing shoe-sized holes drilled in a tree by a pileated woodpecker, he remarked, "This is incredible; your animals are so big here!" Words

Figure 9.9. Beaver dam, with a lodge visible in the background.

I never would have expected from a South African tracker. I suppose true appreciation comes from perspective.

So one theory would posit that beavers have historically played an important role in maintaining high water levels here in this basin. Upon beaver extirpation, the land dried a bit and a forest began to take hold, until the beavers came back to flood the land again. On our aerial map (see fig. 9.3a), at the southwestern edge of the wetland, we see evidence of a dam—though not necessarily that of a beaver—where standing water backs up at the outflow of a meandering stream. Looking closely at the rest of the stream, you can pick out the location of three other smaller dams where the water is backed up—almost certainly made by beavers. But could beavers be the only story behind this wetland?

I often think of beaver-made landscapes as free-form and dynamic. A beaver family will move in and construct a pond, only to eat all their favorite trees and see the ponds fill in. Out of food, the

Figure 9.10. A tree felled by a beaver.

beavers then abandon the pond, the dam breaches, and the trees regrow until other beavers come back a decade or two later. At the larger scale there will be many different patches along the stream, resulting in a shifting mosaic of habitat. At one location there may be a brand new pond. Downstream there may be a pond that was recently breached. Further downstream there may be an old beaver meadow that was a pond fifteen years ago. Such a patchwork of continual change provides great diversity to a landscape.

Looking at the aerial photo, our vast wetland has well-defined edges and extends far beyond the beaver-plugged stream at the south end. This suggests to me that there is something deeper going on here—a static, larger-scale process that defines the basin in which this wetland resides. This isn't just an arbitrary place along a uniform stream where a beaver happened to construct a dam. The shape of the underlying landform seems conducive to holding standing water.

What shaped this basin? Did you notice the big boulders in the wetlands, like the ones on which the *Field Naturalist* students in figure 9.2 are eating lunch? Remember also, as we first entered this landscape, the big boulders scattered in the forest? Where did they come from? These are known as glacial erratics. Chunks of mountains further north that glaciers plucked off and carried southward. To be true erratics these boulders would be made of entirely different rock than the bedrock beneath us—a feature that is useful in charting how glaciers moved. As the glaciers retreated, they left behind rocks of all sizes. Rather than being completely randomly distributed, these glacially deposited rocks are often formed into distinctive patterns—streamlined drumlin hills, sinuous eskers, giant moraine piles, and the dam that created the ancient glacial lake.

If we were to walk through the forest right along the western edge of our wetland, we would encounter dozens of big boulders extending in a north-south line the length of the wetland. Have a look at the topo map (see fig. 9.3b). What does it say to you? It's hard to work out the exact flow of ice and water beneath the glacier 15,000 years ago. But whatever pattern of flow caused this line of boulders, my understanding is that the line of boulders and the underlying mound of till may be what now dams up this basin.

This basic process—of glaciers interrupting the flow of water—is central to the ecology of the region. Staring at aerial photos of Massachusetts, conservationist Matt Burne once circled over 30,000 vernal pools in the state. That didn't include the ones that didn't show up in the photos—the true number is probably double that. The abundance of ponds is, in large part, the result of a landscape that doesn't shed water well. Perhaps water-resistant bedrock helps also.

Where I grew up in Tennessee, beyond the reach of glaciers, we boast that the only natural lake in the entire state is Reelfoot Lake, formed in 1812 when the Mississippi flowed backward into a hole opened up by the New Madrid earthquakes. Maybe you can find other natural bodies of water besides beaver ponds and Reelfoot Lake in Tennessee, but nowhere near the density found in New England. In Tennessee, streams and rivers have been carving up the landscape for millions of years, creating well-developed drainage networks. No matter where you are on the Tennessee landscape, there is almost always a path that will take you downhill to the ocean without interruption, thanks to the long, hard work of water. In New England glaciers repeatedly ravaged water's engineering by arbitrarily dumping vast amounts of unsorted till hither and thither. The salamanders have thanked the glaciers ever since.

In this fen where I am standing, water was perhaps first trapped here when the glaciers retreated. At that time perhaps there was a full-fledged lake before the sphagnum mat grew over it. And perhaps the lake was ringed at one point by the ancestors of the spruce in our opening image. If so, the spruce did not colonize this fen. Rather, this fen is all that's left of the magnificent spruce-dominated forest that used to grow all around us.

MAJOR LESSONS FOR INTERPRETING A LANDSCAPE

- How do wetlands form near your site? What species depend on them?
- Where are the "habitat islands" in your area, and how do organisms disperse among them?
- What ecosystem engineers, like beavers, exert outsized influence on your landscape?
- Does your site show any evidence of sudden shifts in conditions—like a dramatic and sustained change in water level?

Figure 10.1. What you see is what you know.

10. Aliens

INVASIVE SPECIES

Our local university is a monstrous state institution with 30,000 students crammed into less than a square mile of its campus. Manhattan, for comparison, has a population density of 26,000 people per square mile. At our university, in between two dormitory complexes, sits a little, neglected wild area.

Locals call this place "The Orchard," but that name is just a legacy of some long-past history—certainly no visitors to this land in the past couple decades would call it an orchard, although if you search carefully, here and there you can find a half-living apple tree that still bears fruit. More prominently, you'll find isolated eight-foot sections of wooden fences scattered about, and the student oral histories still tell of when the equestrian team used to jump horses over these fences.

Every Sunday morning, while the debaucherous students lingered in bed with hangovers and the bookworm students battled theory in the library stacks, members of the Woodsy Club would converge on this forgotten piece of land to find their place in nature. That was two decades ago.

Standing here today with Juno and Alder, things have changed. But I'm happy to see many familiar faces in the tangled mess be-

Figure 10.2. The Orchard (a) in 1962, and (b) in 2005. The black triangle indicates our location at this chapter's site. (a: USGS; b: Mass-GIS, Sanborn LLC)

fore us (fig. 10.1). We're in the middle of a strange undulating sea of eye-level multiflora rose, bush honeysuckle, Oriental bittersweet, porcelain berry, and privet. The low thicket is punctuated by regularly spaced shaggy forms—short dead trees covered in vines that remind me of Snuffleupagus—that jut about ten feet higher than the surrounding shrubs. In the snow before us, a row of golf-ball-sized dots leading away, alternating slightly from side to side, marks the trail of the animal I had hoped to find here. About one hundred yards ahead, to the west, there's a young forest with lots of aspen in the canopy.

The *Field Naturalist* students stand in a circle under the aspens,

staring down at the handouts I've just given them. A couple of the students fiddle with reddish-brown strips of dogbane bark in their fingers. Their eyes on the sheets of paper, they try to make sense of the maps that depict a series of aerial photographs over the past half century (fig. 10.2). The black-and-white photograph shows that this whole area was in active agriculture in 1962. However, after cultivation was ended, part of the land turned into a forest and part of it turned into a shrubby tangle of invasive species. Why did the two parts end up so different?

Looking at the invasive thicket, I wonder if I'm the only person with a deep fondness for these plants and this place (fig. 10.3).

Figure 10.3. Berries of Oriental bittersweet and other exotics overflow in The Orchard.

All winter long, on the south end of The Orchard, a crew armed with big tractors, cherry pickers, and chainsaws has been hacking away at the plants. It's a heroic attempt to wrest the land back from the grasp of the noxious invasive species and reestablish the native beauty that belongs here. Does anyone else's heart break at the site of this cleanup?

Multiflora rose was historically planted along hedgerows to divide up fields and used to prevent soil erosion. Privet is used for both hedges and flower arrangements. Oriental bittersweet is prized for its decorative orange berries. Porcelain berry sports pastel blue and pink fruits, like little round Easter eggs. All four of these plants originally came from Asia. And all four of them are now despised invaders of the eastern United States.

Usually, when a species is brought from a faraway land, it will die out. Even if it finds a way to survive, it will rarely thrive, much less spread. But with so many species in the world, every once in a while, a new arrival will be perfectly suited to thrive and rapidly overtake a local ecosystem. And it only takes one such species to do major harm.

Why do our native American bittersweet and our native roses occupy modest niches in local ecosystems, while the Asian bittersweet and roses swoop in and take over? Often immigrating species leave behind their enemies—predators, parasites, and competitors—like refugees fleeing persecution. But more like imperial colonists than asylum seekers, invasive species find themselves in the seat of power in their new land, freed from constraints and able to grow relatively unchecked. Some species, like garlic mustard, even bring along new weapons that the other local plants aren't used to dealing with. Garlic mustard engages in chemical warfare by poisoning the soil around its roots—just like the allelopathy our native beech trees use to thwart competitors, but with a whole new set of toxic compounds unfamiliar to our local species.

While non-native plants crowd out native plants, the impacts ripple out through the local food webs. Fewer local insects feed on non-native plants. This, in turn, impacts the bug-eating species, as shown by recent work by Desirée Narongo and her colleagues. In neighborhoods of Washington, DC, that were landscaped with non-native plants, Desirée found that chickadee babies die at significantly higher rates because they are malnourished. Mama and Papa chickadee can't find enough bugs to feed their children in these yards, so the young chickadees starve to death.

Everywhere you go, it seems, the world is threatened by invasive species—and many of the invaders are exported from here. North American beavers are wreaking havoc in Argentina. Eastern US bullfrogs are killing off pond species in the western United States, Europe, and Asia. Eastern US gray squirrels are taking over the habitat of Europe's native red squirrels by first infecting them with deadly poxvirus.

When it comes to invasive species, islands are often particularly vulnerable. Islands tend to be simple, sheltered systems with just a few native species. So new invaders can really flip everything on its head.

Consider New Zealand. The island system has been on its own evolutionary trajectory for millions of years, refusing to participate in the fads of the major continents. There are no native land mammals, except marine mammals and two bat species, which has prompted many birds to trade in their wings, emit strong smells for communication, and approach strangers without fear.

But when Sydne and I visited New Zealand, it was like a giant zoo, harboring all manner of mammals brought by people from all corners of the world. The most abundant wild animals include European hedgehogs, Australian possums, California quail, and Indian mynas. It was a real treat for us to see these weird critters, but they decimate the local flora and fauna. To protect native species, the exotics need to be exterminated. In the United States, as elsewhere, hunters have been key leaders in the conservation movement. Sometimes, however, there's a sense of antipathy between the hunting interests and environmentalists seeking to preserve nature in pristine parks separate from human influence. But in New Zealand it seems the primary goal of most conservation agencies is to kill as many mammals as possible. Shoot, trap, poison, build fences to keep them out. All mammals are pests. Hunting is strongly encouraged in national parks.

The best examples of native ecosystems left in New Zealand are in little habitat islands with intensive management. Often they circle these habitats with big fences to keep invaders out, then they embark on a massive campaign to purge the interior of non-native species.

At the entrance to Trounson Kauri Park on New Zealand's North Island, we found tacked to the wall a piece of paper listing the number of animals killed each year by the Department of Conservation in this two-square-mile park. Over the prior eight years, the figures included 639 cats, 354 short-tailed weasels, 74 least weasels, 21 ferrets, 1,603 hedgehogs, 1,145 rabbits, 1,458 possums, 1,457 rats, 67 blackbirds, and 150 other birds.

If you are both an animal lover and a conservation lover, how do you come to terms with killing for conservation? For the past fifteen years, my cat, Annie, has been my best friend. I can't stomach killing cats. But I can't kill a fly either. For me it all goes back to my existential fear of my own death. I hate killing just as I hate wars. Yet killing is normal in nature. Though I can't picture killing a single chicken, I know that billions of chickens are killed each year for meat. And none of those chickens would have been born in the first place if not for the meat industry. The numbers are so big and theoretical that it's hard to engage with emotionally. It makes me think about the thousands of people who died in the three-day Battle of Gettysburg. That was a huge waste of human life, but then again, that was 1863, so all those people would be dead anyway by now. So how do you score that at the end of the day?

But what are we talking about here? Must we probe into our feelings on life and death? As I was taught in graduate school, conservation biology is a "value-laden science." There is no objective conservation—it is entirely informed by the fundamental morals that guide our personal life choices. To understand how we see a landscape, we must first explore the values that motivate us.

WHALE WRESTLING

From the North Island of New Zealand, it's a three-and-a-half-hour ferry ride to the South Island. It was there that we met the whales (fig. 10.4).

The positive spin is that seventeen of the stranded pilot whales made it back out to sea. But for us the experience was devastating—though there were some moments of beauty. Perhaps we gave

Figure 10.4. A rescue attempt for a pod of pilot whales stranded at Farewell Spit in New Zealand.

them some comfort, and one final swim, but none of the forty individual whales we worked with survived that week. There was mom and baby, who talked to each other continuously. We moved baby next to mom so that they could see each other, and they seemed to appreciate it. In the rising tide baby could swim first and kept wandering a bit away, then coming back to nudge still-stuck mom on the nose, saying "Mom! Come on! Let's go! Why are you still lying there?"

There was the playful teenager who, as we began to swim out, would sidle up to the rescuers, take a breath, then roll upside down for a belly rub from one of us. There was the other energetic youngster who Sydne escorted out but who could really only be controlled by guiding the larger whale that I escorted and to whom the youngster would always excitedly return.

After it was all over, the horrible question that I couldn't get out of my mind was, "Did they shoot mom first or baby first?" Perhaps they had two guns.

Whale rescues are a mixed bag; some are very successful and some end like ours. Before that week, I knew nothing about whale rescues.

Monday night, after leaving the ferry from the North Island, we slept in our car on the side of the road next to a camper van owned by a German military policeman and his uncle. Before going to bed the Germans offered us beers and we chatted for an hour or so. Then the nephew received a text message from his brother back home. On the television screen in Germany, *60 Minutes* was telling the story of a pod of stranded whales that needed help. The whales were only 180 miles from where we sat on the side of the road. We quickly scrapped our plans, found Project Jonah's Facebook page, and prepared ourselves for an early morning four-hour drive to the site of the whale rescue. On the way we stopped by a kayak rental place where they lent us wetsuits for free, without even asking our names.

We arrived on Tuesday afternoon, two days into the rescue effort. On Sunday ninety-nine pilot whales had stranded themselves on Farewell Spit, a fifteen-mile strip of sand that juts out into Golden Bay. Seventeen whales had already refloated themselves on Monday night and headed out to sea, forty-two had died, and forty remained alive on the beach.

New Zealand has more whale strandings than any other country, and every year or so, a big pod lands on this beach. It's a huge intertidal zone, virtually flat for miles. As we were told, once a whale gets in there, it's hard for her to know which direction leads to deeper water and which leads to the beach. During migration, pilot whales travel in big pods and are particularly loath to leave any member behind. If one sick whale ends up on the beach, it can pull the rest of the pod in. Project Jonah members also blamed human sonar, military equipment, and predatory orcas for driving pods onto the beach.

Managing massive whale rescue efforts is the primary purpose of Project Jonah, a well-supported nonprofit with a long history going back to 1974. Within hours of this stranding report, the organization leapt into action. Volunteers from Australia and across New Zealand set down their hamburgers mid-bite, booked airplane tickets, and hopped in their cars to come save these helpless mammals.

There are two primary modes of a whale rescue: keep the whales protected and comfortable while the tide is out, then try to get them floated out to sea when the tide comes in.

When the tide is out, donated sheets protect the animals from a sun that causes black whale skin to quickly burn off and swirl about in puddles below. Pillowcases filled with sand prop animals upright for comfort and so that their blowholes remain above water as the tide comes in. Holes dug in the sand underneath the animals relieve pressure from the two pectoral fins, which are not designed to hold the weight of a whale on land. Finally, buckets of water continuously cool the whales.

As the tide comes in, all the linens come off in a frenzy, and each whale is held by one or a few people. On Farewell Spit the tide comes in extremely fast and cold. Before moving to deeper water, any whales stranded on the outskirts must be carried with slings over to the main pod, which can only be done with the bit of buoyancy provided by the incoming tide. Any sick or badly stranded animals that might hold the pod back have to be shot. When conditions allow, as they did on Monday and Wednesday, one vocal lead whale is strapped to inflatable pontoons. This lead whale is dragged behind a motorboat and is meant to draw the pod in the right direction.

As the incoming tide lifts the whales, each of us holds on to our own whale as best we can and carries her out. When the water gets too deep to stand, we swim as a group arranged in a half circle behind the pod, working with a couple of motorboats to herd the whales like cattle. If all goes well, we follow them out until low tide, some six hours after high tide, and leave them headed in the correct direction in deep water.

Rescuing whales is one of the most strenuous feats I have ever attempted. Waiting all day in the relentless sun and wind for high tide, wet to the bone. Digging your bare fingers through abrasive sand and sharp shells. Listening to the intermingling calls of intelligent whales and overtasked volunteer organizers, all worried and frantic. Drawing motivation from the occasional gunshots that mark the end of another life deemed unsavable. Watching stressed whales flail, bleed, vomit, and die. Moving limp, two-ton animals on dry land. Trying not to inhale the smells of the whale breath and shit.

Then when you can stand and they can float, you spend a couple hours wrestling the whales in cold water. When the water gets above your shoulders, you spend a couple hours trying to outswim the whales. When you can no longer control your shivering, you take a motorboat to shore where you sit amid the smells and sounds and bloody rain of rotting whale corpses violently exploding from internal pressure buildup. When you warm up, the boat takes you back out, this time two miles into the bay, where they drop you off to chase the whales through frigid water some more. When complete darkness falls, you return to your tent and get up at dawn to do it again.

Yet to refloat a pod of whales is one of the most beautiful things I've ever done. You become part of the pod. The tide rushing past your ankles brings excited anticipation to you and the whales, as both species begin to talk and fidget more. Thirty minutes of rising tide fills a desperate situation with hope as huge, lifeless black lumps turn into graceful gliders. A body that has been pinned prostrate for twelve hours now rolls intently from side to side and upside down, weightless and frictionless. Friends reunite against all odds from opposite sides of the pod. In deep water the sun shimmers off many receding black fins, and people hold hands in a circle around them. Then the shapes drift out of sight, slipping into an unseen life that you have touched.

Everything went perfectly on Wednesday night. Exactly why they were stranded again by Thursday morning, we don't know. But it was the sixth time the whales had been stranded at low tide in the past four days, and they were worn out. Some of them may have been the same whales that were stuck nearby two weeks earlier. We were working well into a world-renowned bird breeding habitat where public access is usually restricted. We had expended enormous amounts of physical, emotional, and financial efforts for the cause. The last few days had entailed hundreds of volunteers, many Department of Conservation officials, a couple large beach buses, an industrial digger, and lots of donated sheets and food. For the sake of all involved, the decision was made to end the effort. At least the seventeen survived.

SAVING SALAMANDERS

Another, smaller-scale rescue operation takes place each year a mile and a half northeast of The Orchard. Spotted salamanders live most of their lives underground in mammal burrows in upland forests. But on the first warm, rainy night of the year, they emerge and migrate hundreds of yards down to their breeding ponds. They mate for a couple weeks, then march right back up to their home burrows. But if people break the connection between the pond and the uplands, it spells trouble for the salamanders.

At this particular site there's a road. More to the point, there is a regular onslaught of tires that squish salamanders on the road. So every spring for years, the local children and nature lovers

Figure 10.5. A spotted salamander rescued from the road.

would come out to this site to direct traffic and help the salamanders across the road. The "bucket brigade" would walk along with flashlights, scooping up nine-inch-long black and yellow salamanders from one side of the road and dumping them on the other.

Then researchers installed a set of shoebox-wide tunnels under the road for the salamanders to use. Importantly, they installed fences along the roadsides that guide the salamanders into the tunnels. For the most part these tunnels are a big success. Still, enthusiasts converge every year to watch and to help the occasional individual who jumps the fence. It's a wonderful point of connection between the community and nature (fig. 10.5).

But it's not like spotted salamanders are particularly rare. There are tens of thousands of spotted salamander populations in the state. On the drive to participate in the annual salamander crossing festivities, I and others roll our tires over many other salamander crossing spots that don't happen to be famous.

With the salamanders and with the whales, we become emotionally entangled with the individuals whose lives are on the line. At some point it's not so much about the species as a whole but about our relationships with the animals. And it's about saving the things we know and love.

ANIMAL FORMS

For me, The Orchard is a place I know deeply. As I look up at that little glacially sculpted hill in the distance, I see myself emerging from a debris hut constructed of sticks and leaves one October morning. The night had been a bit scratchy, but I woke up toasty and surprised to see snow on the ground as I pushed my way out of the leaf door.

For ten winters The Orchard was the destination for the first field trip in the annual animal tracking course that Charley and I taught. Why here? Because there's a lot going on.

I park the van in the lot by The Orchard, and our animal tracking students spill out. There's a fresh layer of snow on the ground, and Charley and I lead the class to a spot where the snow is undisturbed. Before we look at footprints, we first have to teach about animal gaits—how animals move.

I get down on all fours and pretend to be a raccoon. Pausing frequently, I lean onto my back legs, dig for frogs in the mud with my front hands, and munch on the morsel. Then I keep walking. Some students guess the species correctly, then we look at the track pattern in the snow. The tracks of my left rear foot land next to my right front foot, but slightly offset. Likewise, my right rear and left front feet are paired. This is an unmistakable pattern that virtually always signifies raccoon. Driving the van over a bridge at forty miles an hour, you could spot the remnants of a month-old, melted-out raccoon trail on the ice below and be completely confident of the species.

Again walking on all fours, I then spread my weight evenly between my front and back legs and creep deliberately forward. At several points I pause mid-step with one foot in the air and three feet on the ground. Then I push on, keeping my head level the entire time. As I glide forward, I carefully slip my right rear foot into the track of my right front foot and the same on my left side.

Some students give each other sideways looks of concern about my sanity, while others eagerly guess that I'm trying to be a cat. Inspecting my footprints, we see the tracks evenly spaced along the length of the trail, alternating back and forth from left to right. This is a direct-register walk, where the back foot lands directly on top of the front foot tracks. I circle three of my tracks and explain that, when an animal is in a walking gait, circling three tracks gives you an approximate size of the animal.

Next we demonstrate the "trot." Standing on your right leg, you put your left leg and right hand out in front, with your left hand by your side. Then, all at once, you switch sides. Hopping onto your left leg, you put your right leg and left hand out in front. The whole class dances to this beat, then I get down on all fours next to my walking cat trail. I hesitate for a minute wondering if this is really a good idea. I start by trotting in place—standing on my left front and right rear feet, then hopping to the opposites. After a few such switches, I add forward momentum and trot out through the snow. I try to picture a dog bouncing casually across a field next to her owner. I stop to pee on Charley's leg and trot on.

Inspecting my trotting trail, we see that it is essentially the same pattern as the walking trail—my rear foot landed directly in the track of my front foot on each side. The tracks are evenly spaced along the trail, alternating left and right.

I call a volunteer student and have her walk on her two legs, like a normal human. Then I have her come back to the starting point and run as fast as she can next to her walking trail. How do the trails differ? One student points out that the running trail is messier—the snow is exploded out of each track. True, but what about the pattern of track spacing? Another student says the running tracks are further apart.

"In which direction?"

"Well, the direction she was running."

"Yes, the strides are longer. What else?"

Awkward silence.

"What if you look from the end down the length of the trail?"

"The walking one is further apart."

I bend down and, quietly drawing in the snow, trace lines along the edges of the two trails, showing the trail width—the straddle. The students now see that when she ran, she stretched her legs out in longer strides, and this resulted in a narrower trail width. Looking back at my walking cat and trotting dog trails, we see the same features. In the trotting trail the strides are longer and the straddle is narrower.

At this point I break open a backpack full of old gloves and ask all the students to walk and trot on all fours. Most oblige, and laughter erupts as a dozen adult humans tumble around in the snow trying to place their front and hind feet in a line.

Back in our circle, having covered the "alternating gaits," where the left and right sides of the body take turns, we move on to the more exciting "bounding gaits," where the animal throws her whole body through the air. Here, the front legs work together as a team, as do the back legs.

Channeling my inner gray squirrel, I squat in the snow and apologize to my wrists. Then I leap off of my back legs, reaching before me with my front legs. My front legs come down first, but only for an instant. Just as my rear legs are swinging down, my front feet spring off, letting my rear feet slip past into the snow. My rear knees bend and launch me again flailing through the air. After five such bounds I crash ungracefully.

In my trail we see my tracks arranged in little clusters of four. Within each cluster my two front feet are next to each other and in the back of the group. My two rear feet are wider apart and toward the front of the group—remember they swung past my front feet as they landed.

The rabbit trail I create is much like that of the squirrel, except my front feet are more in line and staggered with one far in front of the other. Whereas the wide stance of a squirrel is designed for climbing trees, rabbits are built for making sharp turns around bushes while dodging bobcats.

I move on, attempting more difficult bounding gaits with diminishing success but growing laughter—loping fox, galloping dog, and, finally, the weasel, to which I attribute chronic back pain.

Why are we crawling around like animals? Tracking students come expecting to learn the shapes of footprints. They want to hear that dog feet are overall long and narrow because they are built for forward movement, that dog feet are symmetrical, that dog tracks display big meaty claws, that most of a dog's weight is forward on her big toe pads, and that the negative space in a dog's footprint forms an "X," as in, "X marks the Spot." Cats, built more for balance, have wider feet with toe pads that are much smaller than their heel pads, claws that rarely register, and asymmetric tracks wherein there is one leading toe that juts forward. Extending my middle finger to the students, I offer the mnemonic "Cats have catitude." Such generalizations work pretty well to divide species into appropriate taxonomic groups, from cats to dogs to weasels to rodents and so on.

But looking at individual footprints is just a small piece of what Charley and I do in tracking. The overall trail pattern tells you a lot of information, and sometimes, such as when the tracks are old or melted, it's the only information you have. Like foot morphology, the way that animals move tends to depend on what taxonomic group they belong to. Knowing the base gait of each animal, and then recognizing this trail pattern, helps you identify the species. Of course, cats can trot, dogs can walk, and everyone can gallop when they want to. But those aren't the most comfortable ways for them to move. If an animal is doing something other than its natural gait, you've got to stop and ask why.

VOLEY-VOLE

Charley and I lead the students along a trail that separates the aspen forest from the overgrown invasive tangle, heading toward a grassy field on the north end of The Orchard. By the edge of that field are some large shagbark hickories that I remember from one of my first Woodsy Club adventures. It was September, and we sat gathering hickory nuts, cracking them open, and devouring the delicious flesh inside. It was the first time I'd just been out in nature foraging for nuts.

In the middle of the field, there's a large patch of dogbane that we later gathered to make natural rope. The inner bark of dogbane is one of the strongest plant fibers you can find, and, twisting it in a reverse-wrapping technique, you can quickly make a remarkably strong rope. I once gave a ⅛-inch thick dogbane string to my roommate in college who was much stronger than me and asked him to break it with his hands. Grabbing one end in each hand, he strained his muscles, clenched his jaw, blew air out his mouth, puffed up his cheeks, and turned his face bright red. The string held.

When Charley and I arrive at the dogbane patch with our class, the snow is gone from much of the field, leaving the grass exposed. In the middle of the field, we find a fresh dig in the grasses with a tiny scrap of dark, mangled fur next to it. Sniffing the grass on top of the dig, I inhale a strong skunky smell—red fox pee. Here a fox had stood patiently listening until a vole, running through a tunnel in the grass below, was in just the right position. The fox pounced and caught the vole, then left a spot of urine to claim the hunting grounds.

Prior to colonization, red foxes in North America were a boreal species, living mostly in Canada and mountains of the West. In the 1600s, English people obsessed with foxhunting brought over European foxes and introduced them to the Colonies. These introduced foxes proved to be generalists and spread throughout the continent, displacing our native gray foxes and the native populations of red foxes.

When Charley and I bring students into The Orchard fields, they want to track the foxes. But Charley and I want to show them the voles. The voles are the reason the foxes come here. We bend down and part the grasses with our hands. Within seconds, little vole tunnels appear. We tell the students to go find their own vole tunnels. Each trail is about one inch wide, edged with clipped grasses, and decorated with rice-grain-sized scats that fade from green to black.

Voles, sometimes referred to as field mice, are the basis of the meadow food chain (fig. 10.6). Foxes, coyotes, weasels, and raptors are all here to eat them. But they're often overlooked by humans.

One fall, Charley and I were studying for an upcoming wildlife tracker evaluation, and we needed to work on distinguishing vole tracks from those of white-footed mice. Looking at gaits can be helpful in telling voles from mice. Mice, like squirrels, are built for bounding along and climbing. Voles are built for trotting along through short tunnels. If voles were to bound through their tunnels, they would keep hitting their heads on the tunnel ceiling. By trotting instead of bounding, voles can keep a more consistent height. Vole tracks crossing an expanse of mud or snow will often show this preference clearly. You'll see an evenly spaced row of

Figure 10.6. A juvenile
meadow vole.

footprints alternating between left and right. Mouse tracks crossing through the open, on the other hand, usually show up as little clusters where all four feet land and then launch the animal
through the air again.

But sometimes it's helpful to know how the shape of individual
mouse and vole footprints differ. So Charley and I set out to catch
a vole. We borrowed some small mammal traps from a wildlife researcher in my department and scattered them across the fields of
the university. Every day, Charley and I checked the traps. Every
day, they were empty. After a week of this, we gave up. I collected
the traps and threw them in the passenger seat of my car.

That evening I got in my car and headed north to the small
cabin where I lived fifteen miles away. Halfway through my trip,
driving on a two-lane state highway, I saw one. There, shimmer-
ing in a light rain, scurrying around on the black road surface, il-
luminated by my headlights, was a little, gray, vole. On the side of
the road was a small grassy shoulder in front of a big patch of gold-

enrod in someone's yard. I swerved onto the shoulder and threw my car into park. Grabbing one of the traps from the passenger seat, I leapt out of the car and dashed over to the vole. It just sat there, staring at my headlights. I bent over and gingerly pinched the scruff of its neck. I picked it up and gently dropped the vole into the trap.

"Voley-Vole," as she came to be known, lived in a ten-gallon glass aquarium in my cabin. We ran her through sand, mud, and flour. We studied the thin lines of her toes. In our collection of animal track plaster casts, numbers 262, 263, and 264 still show her footprints. Annie liked to sit on top of her cage looking down. I fed her garlic mustard from my yard, along with bird seeds and guinea pig food.

Once we understood Voley-Vole's footprints, I intended to release her where I had found her. But winter set in, and I felt bad dumping her in the snow with no home or food. So she stayed through winter. When spring came, I was swamped with classes and kept forgetting to take her on my daily commute past the spot where I'd found her. Eventually, in a spring cleaning frenzy, I decided to just take action and dump her in the yard at my cabin. Maybe, Charley likes to say, Voley-Vole was now trained to eat garlic mustard and would help cut back its invasion into that forest.

Six months later we bought a house. In front of that house was a big patch of goldenrod near a grassy road shoulder. I didn't realize it right away, but that house is where Voley-Vole came from. We had traded places. I now lived with Voley-Vole's family, and, so it seems, they were hell-bent on revenge.

A set of healthy apple trees grew in the yard of our new house and gave us delicious fruits the following summer. A year after we moved in, working beneath the winter snow, the voles girdled all our apple trees. In spring I found the lower half-foot of the trunks completely stripped of outer and inner bark. With no green grass in winter, apple trees are a great snack for the tiny herbivores.

Figure 10.7. Elm tree girdled by voles.

Plus tree bark has sodium and other minerals that the voles are deprived of in winter. Other herbivores, like moose and deer, have the same winter craving for bark, and their sign on trees is even more dramatic. Our apple trees all died. One of them even had its roots so chewed up that I was able to lift it out of the ground with one hand.

Voley-Vole's family didn't stop at the apples. They girdled and killed an elm that was eight inches in diameter (fig. 10.7). They girdled a chestnut tree that we had planted. They chewed our garden hoses to bits. We planted new apple trees and surrounded them

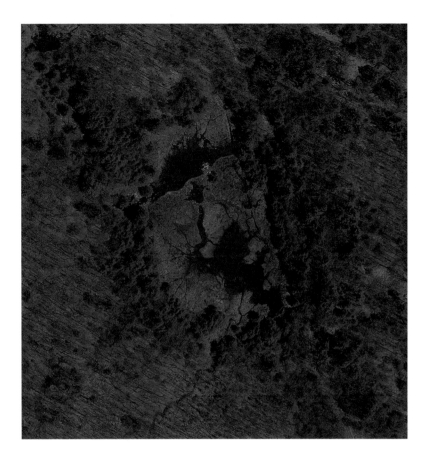

Figure 10.8. A ring of pines surrounds the beaver meadow where the coyote barfed-up voles. (Google Earth)

with two-foot-high hardware cloth to keep the voles out. The following winter the snow accumulated to about two and a half feet, and the voles found a route over the top of the barrier. The following spring we found our new apple trees girdled high off the ground.

The girdling of our elms and chestnuts surprised me because I thought voles wouldn't like them. Apparently I was wrong. Which trees voles choose to munch on is important. It can determine, for instance, the course of forest succession. Many trees, like aspens, oaks, and pines, have developed complex chemical defenses to ward off would-be herbivores. In a field full of voles, pine saplings will survive while ash and maple saplings will get gobbled up.

Researchers working with beavers and quaking aspen have shown that as the beavers cut back shoots, the trees will respond by producing even more beaver-repellent chemicals in the subsequent shoots they grow. At the landscape level beavers show how herbivore preferences can alter forest composition (fig. 10.8). In the forest around the edge of the meadow where I climbed into the beaver lodge, there's an abundance of pines, unlike in the forest a few steps beyond. You'll find many old beaver ponds encircled by such rings of pine. Venturing only a short distance from the pond, the beavers selectively ate the choice hardwoods and left the distasteful pines to grow in this zone.

ALARMS

As our tracking class walks back from the field where the dogbane grows, we pass the weeping willow under which Woodsies used to gather making willow baskets and eating wild grapes. These are fox grapes—the wild precursor to Concord grape and other cultivars. In the forest behind my house, some of these grow giant, arm-thick vines that stretch up into the tall canopy. They're too high for me to taste the grapes, although, in a thirsty pinch, I could cut into these vines as a quick source for water—I've heard big ones will sometimes pour water out like a faucet when cut. Those vines behind my house puzzle me. I think of grapes as early successional species. So do the vines date back to when the forest looked more like The Orchard? Did they grow up along with the trees? It would be interesting to count the growth rings on lots of thick old vines and their associated trees. Well, those grapes are dead now. A couple years ago our neighbor severed the grape vines at the bases, presumably displeased at their tangled appearance and seeing them as parasites of the trees. But I love the tangled appearance: the fruit feed wildlife, and each vine seems to have struck a long-lasting relationship with the tree it depends on to reach the sun. I wonder, did these vines gush water when my neighbor cut them?

In The Orchard, near the still-living grapes, the tracking students notice a big flock of robins. "Don't they fly south for the winter?" "Yes," Charley answers. Some robins do, but not all head south to Florida. Many just stick around where they can find a good winter food source. The Orchard is overflowing with tasty berry-producing plants like multiflora rose, privet, and bittersweet.

I pluck a rose hip off the end of a thorny bush, pop it in my mouth, and suggest that the students do the same. I carefully separate the thin, red, sweet, tangy, vitamin-C-filled skin from the bulky wood-like seeds inside. I spit the seeds onto the ground. Someone once told us that if you chew up the rose seeds, it will make your butt itch. Because of that, Juno now likes to chew and swallow the seeds.

Amid the loud chattering of robins, Charley launches into a lesson on bird language. At the most basic level, if you listen to the birds, they will often warn you of predators. If you learn to read them, you can tell what and where the predator is. In college, while I was trotting barefooted behind a red fox in the woods, several blue jays joined the parade, noisily dive-bombing the fox as we went. This summer we learned to follow the gray foxes through our meadow by listening to the family of house wrens fledged from Juno's birdbox. While Charley is explaining bird language to our class, as if on cue, the robins and all other birds suddenly get miraculously quiet. Are they responding to us? We wait for several seconds, then a Cooper's hawk glides overhead.

One evening this summer, while doing our nightly tick check on the kids, I heard robins chirping loudly outside. Then I noticed a chorus of other species join in. I popped the screen off our second-floor bathroom window, and Juno and I leaned out. Back in the forest, about mid-canopy, woodpeckers, thrushes, and various songbirds were excitedly flitting about and screaming. Then the alarm cloud shifted to another position in the forest, still about halfway up the trees. I knew exactly what species they were after. I told Juno to stare at the spot in the woods with all his concentration and wait. After a few minutes we caught a glimpse of the barred owl flying away through the trees.

In that forest behind our house is a network of trails—mostly wide logging roads used by ATVs, the occasional pickup truck, and sometimes logging skidders. Recently, Annie has been following us on long hikes there. On Annie's first hike, we reached the crest of the first hill where another trail splits off. Annie, who had been walking down the center of the trail the whole time, approached a knee-high hemlock sapling that leaned in from the edge. As she sniffed it, her ears turned backward, her spine stiffened, and her fur stood up a bit. Then she crept off to the opposite side of the trail and walked into the woods a couple feet. The hemlock looked to me like a classic place for a bobcat to spray, and a quick sniff confirmed my suspicion. As we continued our hike, Annie, now aware that she was not the top cat in those woods, walked in the forest parallel to the trail a few feet from the edge.

Months later, on a 90-degree May day, Sydne and I returned home a bit before sunset to find Hannah Harvester, our nanny, walking down the trail behind our house with Juno and Alder, frazzled. They had walked the mile up to The Crevice, and Annie, kidney failure and heart disease be damned, had followed them the whole way. The whole way, that is, until about ten minutes before they returned home. Right around the first trail intersection, she had disappeared. They tried but couldn't find her. Annie had never been up that route before, and she's not one to wander far alone—her instinct if left alone in unfamiliar territory is to hunker in place until I rescue her. I dashed up to find her, calling her name as I went, in part to get her attention and in part to ward off any wandering coyotes or fishers. If this was going to be Annie's end, at least she was outside, which is where she's happiest.

When I got to where Annie was last seen, I stopped and listened. All was quiet. Except, in one particular direction, there was a subtle, multispecies alarm call coming from within the forest. A few birds and a chipmunk were softly chirping. Their energy seemed to be directed at the ground, at a spot perhaps fifty yards away. Because of the way the trail turns, the alarm call was actually on a straight-line path back to the house, even though it was far off-trail. I headed to the spot and found, perched on top of a cliff, sprawled out beneath a hemlock, an exhausted, but satisfied, calico.

BROWSE

A few minutes after the hawk passes over our animal tracking class in The Orchard, the robins, cardinals, and other birds resume their various conversations. With my back to the aspen forest, I lean over into the rose thicket on the other side of the trail and point to where one of the stems has been clipped. All that's left of the thin stem is a little stub, cut sharply at a 45-degree angle. Who did this?

Deer don't have upper incisors in the front of their mouths. When they bite things, they leave behind ragged tears. It's the rodents and rabbits, with their sharp incisors, that make such clean, angled cuts. Like on the hemlock twigs nipped by large, quill-covered rodents at The Crevice. In The Orchard it's rabbits, not porcupines. Rabbits love multiflora rose. We call such thickets "rabbitat." I ask all the students to spread out, crouch down, and find their own examples of where rabbits have clipped rose stems. Within a minute everyone starts finding rabbit browse.

Eastern cottontail browse, to be precise. If we were in a dense thicket in the understory of a forest, I'd expect to find snowshoe hares. But here among open fields and shrubs, it'll be cottontails. Historically, it actually would have been New England cottontails that lived here, but today they are threatened with extinction.

After farms were abandoned and early successional thickets grew in, New England cottontails rejoiced at all the new habitat. In the early 1900s there were more New England cottontails than ever before, but the celebration was short-lived. As forest succession proceeded, the rabbit-friendly habitats were shaded out by trees, and the populations crashed. Conservationists, concerned with the lack of rabbits to hunt, imported over 200,000 eastern cottontails from the Midwest to bulk up the New England populations. Eastern cottontails had never before lived here, and they brought a secret weapon: their eyes.

A pair of eastern cottontails and a pair of New England cottontails are foraging side-by-side under the safety of dense shrubs. While brown pellets—the size and shape of M&Ms—accumulate around snipped twigs, the last of the good food is depleted under the safety of the shrub. However, there is still some food out in the open. One of the eastern cottontails, whose genes were trained in the open country of the Midwest, hops out away from the cover. One of the New England cottontails nervously considers for a moment, then stiffens her spine and joins her in hopes of better food.

A coyote, out doing her nightly rounds, trots onto the scene and spots the foraging rabbits. The eastern cottontail, whose eyes have about 40 percent more exposed surface than those of the New England cottontail, sees the coyote first and dashes for safety. The small-eyed New England cottontail doesn't see the danger until it's too late.

Back under the shrub, the remaining New England cottontail faces a choice: go out with the bold eastern cottontails to forage in the open where she just watched a coyote eat her sister and she knows there's still a hungry red fox nearby, or stay under the shrub and starve.

In this way, patch by patch, the slightly larger and more versatile eastern cottontails have thrived while the New England cottontails have died off.

COYOTES

One Sunday afternoon in October of my junior year of college, I was driving back from my grandparents' house, where I had been visiting my dad. It had been a month since he was diagnosed with pancreatic cancer, and, on my drive back to college, I knew I needed to find my Woodsy community for support. I would be late for the fire in The Orchard, where the plan was to boil Jerusalem artichokes with bear grease. I was still on the interstate with about twenty miles to go when I saw a dead coyote lying next to the metal guardrail.

The eastern coyote is a modern invention. Historically, coyotes were a western animal. In the East we just had wolves. But then we killed all our wolves, and coyotes marched eastward to fill in the newly opened niche. As they moved in, the invading coyotes interbred with some of the remnant wolves on the edges. They finally arrived in this area in the 1950s. Fortified by wolf DNA, these new eastern coyotes are much bigger and more prone to working in packs than western coyotes.

When I reached The Orchard that Sunday, we decided to add "coyote skinning" to our plan for the day. One Woodsy, Mary, came with me to retrieve the animal. We first stopped by my college where I, always the hypochondriac, raided a janitor's closet for plastic bags and gloves to protect us from any coyote-borne illnesses. Retracing my drive, we arrived at the coyote, and I pulled off onto the side of the interstate. As I set my hand into the fur on her side, I could feel the coyote's warmth still radiating. Mary and I lifted her up and set her on some plastic bags in the trunk of my gray Volvo. Somberly, we headed back to The Orchard as the sun dropped in the sky.

The rest of the Woodsies sat around a campfire in the forest up on the small hill at the east end of The Orchard. We laid the coyote down in the dry leaves near the fire. For a while we sat in silence just admiring the beautiful animal. Sunlight filtered through the trees onto her patches of gray, brown, reddish, and white fur. After paying our respects we got to work.

We carefully cut skin away from muscle, peeling back her outer layers. As the skin came off her legs, I held on to a foot and moved it in a running motion. We watched in awe as the ligaments and muscles, now unmasked, retracted and extended before our eyes. We skinned the coyote late into the night. The palpable emotional intensity grew throughout the process. It was no longer a coyote; it was one of us lying there in the leaves with our skin coming off. Our inner workings were being exposed for the first time. At the final stage we suspended the skinless body from a tree, head down. Her inverted skin, still attached at the nose, hung like a mirror image looking upward. I wondered what would happen if any of the other 30,000 university affiliates had wandered into this late-night scene, where a band of humans stood in the light of a campfire around the gruesome spectacle of two inside-out dogs attached at the nose hanging from a tree.

Her feet were too difficult to skin, so we hacked off the bottom six inches of her legs, and I later placed them in salt to cure—I wanted to keep them to study her tracks. The feet quickly took on a rotting smell, so I hung them in a white plastic crate outside my second-story dormitory window. During my last night on campus that semester, I was up late cleaning my room before an early morning trip home for winter break. I decided it was time to part with some of the larger rocks on the floor of my room. The window above my bed pivoted sideways to open and seemed like a great place to eject the rocks. But the window was closed. In a flash of brilliance, I invented a game where I would stand in the middle of my bedroom and gently toss a rock toward the closed window. Then, racing the rock, I would leap up onto my bed and open the window before the rock got there.

I won the first race, and the rock went sailing out of my open bedroom window. I won the second race too—but there was one problem. When I had closed the window after the first race, the latch had slipped down. So, although my hand reached the window before the rock, the window didn't open properly. The metal frame stayed fixed while my hand and the rock went through the glass together. Big shards of glass and one rock rained down into the snow, as my wrist gushed blood.

I wrapped my bloody hand in a towel and secured it with a rope. After a trip to the emergency room, I called campus facilities to have them come fix my window during winter break. It was then that I remembered the coyote feet hanging in the crate outside the window. I really didn't want college officials finding the coyote feet—I was already flagged for climbing on the roof of the gym, for searching out the campus steam tunnels by prying up manhole covers with a crowbar, and for the lockpick set that arrived in the campus mailroom in a box marked "guns and ammo." So in the wee hours of the morning, with one arm wrapped in bandages, I carted my crate of coyote feet far into the woods behind the college and lashed them to a tree. After winter break I eventually moved the feet to a different tree where they stayed for years, slowly deteriorating.

The coyote's carcass sat for a while in a Woodsy's basement freezer, but before we could eat it, a plumber searching for an outlet unplugged the freezer and forgot to plug it back in. Three days later there was a warm freezer full of spoiled meat. The coyote skin was passed back and forth among Woodsies and now sits in a plastic storage tub with other Woodsy memories in the back of my house.

TRAILCAM

In writing this chapter I decided I wanted to have pictures of the actual individual animals that live in The Orchard. Really, there's one species in particular that I'm after—the predator that typifies this habitat. I have a few old shots from a decade ago when I disassembled a film camera and a motion-sensing chime and wired the two together. But I'd like some better images. So this winter I ordered a cheap digital trail camera online, set it up in The Orchard, and set out to check it every week or so.

When we first visit The Orchard this winter, Juno is excited to find and collect sumac berries. Staghorn sumac is one of the few native plants that grows in the tangle. Its big head of red berries, when dunked in water, makes a delightful lemony beverage—high in vitamin C—that we call "sumaconade." The sumac shrubs where I used to collect berries on the south end of The Orchard have been shaded out. We still see the stems, covered in fuzz, resembling velvet-covered deer antlers, but no fruit.

We walk further north and spot a bunch of red berries several feet into a thicket of multiflora rose. The cluster of fruit sits atop spindly stems as if drawn by Dr. Seuss. I fight my way through the sharp rose thorns that grab my clothes and hold me back with their backward curve. Thwarted, I return to the trail and try another route in. Again, the thorns keep me at bay. After one last attempt I retreat and we look for berries elsewhere. The next bunch we see is at the top of a fifteen-foot sumac tree. Too high. At last we find some at just the right height and near the trail. We snap off a few clusters, plunge them into our water bottles, and enjoy instant sumaconade.

On Christmas day it snows six inches. After eating out for Chinese, we go to check the trailcam. Sydne and Alder fall asleep in the car, so Juno and I walk into The Orchard to dig the camera out of the snow. We're the first animals out after the storm, and it's a lot of work for Juno's little legs to plow through the snow.

About fifty feet from the camera, it gets really hard for Juno to walk. I forge a trail ahead of him, and he walks behind, holding my hand. I take very short strides, punching holes just the right distance for Juno's feet to glide into as mine lift out. Coyotes will do the same thing for each other to save energy. Often, when following what looks like one perfect coyote trail, the trail will fan out into the trails of several different animals that were all stepping exactly in each other's tracks. Juno beams as we walk toward the camera, "Dadda, you and me are coyotes!" I'm big coyote and he's little coyote.

The camera flashes at us as I bend down to pop the memory card out. I set my laptop in the snow and transfer the images. We expect to find the picture of ourselves as we approached the camera. Instead, all we see on the screen is a big, beautiful coyote. On some level, we truly believe, as documented by the camera, that we have actually transformed into coyotes. Juno and I howl the whole way back to the car.

The next time I return to check the camera, I bring both Juno and Alder. While we wait for my computer to boot up, we attend to another mission. All winter we've been trapping mice in our house, using the same little trap that once held Voley-Vole. We don't have the heart to kill the mice, but we don't like them in our house either. So we dump them into a little ten-gallon aquarium, along with scraps of leftover human food, cat food, and bird food.

When we caught the first mouse, it was well below zero out, and I figured she would freeze to death in the cold. So we held on to her for a while. Juno and I constructed a little cardboard house for the mouse with multiple rooms, nesting material, and food stor-

Figure 10.9. White-footed mice in temporary housing.

age. The triangular silhouette of the cardboard house vaguely resembled the real house we live in. We transported the mouse in this house out to The Orchard and let her go.

After the first mouse, just to be safe, I set the trap again to see if there were any more. There were. It turns out that a single mouse nest can have dozens of individuals in it. That's how they keep each other warm in winter. I wondered if the first mouse, even with its extravagant house, was able to keep herself warm after we moved her to The Orchard. The next triangular house had two mice in it, and we placed it in the same spot in The Orchard next to the first mouse. I figured maybe the three mice would find each other again. The next house had four mice in it, and it was so cute to watch them huddled together in a line on top of the house inside the aquarium at night (fig. 10.9). Are they so different from pilot whales? I soon lost count, but we've now delivered over twenty mice to The Orchard.

We place our latest mouse house in the growing pile and return to the computer to download the images from the trailcam. The images show rabbits, gray fox, deer, joggers, and mice, possibly from our house, but not the one species I'm after.

By mid-winter we now have two cameras to check on, but one of them is aimed at our mailbox. A few weeks ago the kids and I built a new mailbox for our house to withstand the plow. Out in front of the "Wildlife Habitat" sign, the mailbox looks like a life-sized black bear. We named it Ursula. But in the night someone's been intentionally ramming it with their plow, so I pointed a camera at it, programmed to catch any nighttime activity.

TRAILING CATS

On the last field trip of our animal tracking class, Charley and I park the van at the entrance to a huge forested area. After about a half hour, we pick up the trail of a bobcat. The students are elated, and we decide to spend the day trailing the cat. We follow as it walks through the open forest. Walking at the rear of the gaggle of students, Charley and I notice where the trail pattern changes in front of an old rotting stump. But the students, now far up ahead, have breezed on by this stump in their excitement at chasing down the trail.

We call the students back to smell the fresh pee on the stump. Always pay attention to changes in the trail pattern, and never pass up an opportunity to smell pee when you know exactly who made it. Basically like house cat pee, this scent always reminds me of the sickly sweet smell of someone's basement who has too many cats. We continue following the cat as it skirts the edge of a beaver pond, where we stop for lunch. After lunch we again pick up the trail, following the cat back into the forest. Then Charley and I spot a dense thicket of white pines in the distance, and we give each other a knowing look.

If you follow a bobcat for long enough, it will inevitably take you to a horrible thicket, be it mountain laurel, young white pine, blackberry, or multiflora rose. This is where they hunt. In this particular case we expect snowshoe hares to be on the menu. Inside the thicket should be a rectangular impression in the snow where the cat sat, waiting for lunch to be served. A hunting lay is a great track for students to see. But the rabbits and the cat love the thicket because it is virtually impenetrable for larger species. Twelve humans breaking their way through the pines without losing the trail or destroying it seems unlikely. Eager to experience the world through the eyes of a bobcat, the students follow in. Charley and I walk around to the other side of the thicket and wait.

On another day of our animal tracking course, we trail an opossum at The Orchard. We follow the trail to a hole in the ground. I recount the experience of waiting by the little cave behind my house in Nashville just after sunset and having an opossum nearly trip over me as it darted out of the hole and along the top of the log I was sitting on. Charley and I decide to come back that evening and wait for the opossum.

Opossums aren't built to survive a New England winter, and they never used to live here. Then they discovered shelter in the warmth of houses and the extra food in people's garbage cans. Only by relying on humans can opossums survive the winters. In the summer they fan out across the landscape. But those that don't come back to humans before the following winter will die. In this way they have been steadily marching northward.

When the sun goes down, Charley and I return to the hole at The Orchard to wait. After a half hour, a dark shape emerges from the thicket a hundred feet west of us. It's a bobcat. It walks up the bank of a road and out into the middle of the traffic lane, right in front of a large public transit bus, which the cat casually ignores. Under state law, drivers are required to stop for pedestrians crossing the road, and the cat's self-assured stance implies that she knows this. The bus driver brakes and the cat pauses, shoots a long look over at us, then ambles on to the other side of the road into a forest of oaks, maples, and ash.

Bobcats. They're the main reason we bring tracking students to The Orchard. It's the only place I know where you can always find their sign. Why? Because the big, tangled thicket stocked full of rabbits is bobcat heaven. For the past twenty years, The Orchard has consistently held the densest concentration of bobcats of any place we've visited. On full-day field trips out to expansive forests, we might expect to see bobcat sign every three or so days. In The Orchard we always expect to find bobcat sign within the first half hour.

It's not that there are a lot of different bobcats here, it's just that the one family of cats doesn't wander far. Out in the forest a bobcat may prowl a territory that covers many square miles. But where there's unlimited cover and food, why bother going anywhere?

MOVING THE CAMERA

The second week in January, I return with Alder to The Orchard. We've just dropped Juno off at school, picked up new batteries for the camera, and left a prescription at CVS. We have about fifteen minutes before the prescription will be ready for pickup. I figure it shouldn't take long to swap out the batteries. The problem with the batteries is that the camera is pointed west. As the sun sets, it shines into the infrared sensor and triggers it continually, eating up the power.

I'm superstitiously hopeful that we'll have caught a bobcat—so that the universe can repay us the balance from yesterday's tragedy at our mailbox. At around 8:00 AM on Sunday morning, someone stole our entire bear, mailbox and all. I'd set the stupid trailcam on our driveway to stop triggering at 7:00 AM. The camera had watched the entire federal crime unfold but didn't capture a single image. Devastated, the kids and I are determined to build a new bear. This time I'm going to hide a GPS tracker inside.

At 15°F, it's a frigid morning. I nestle Alder into the carrier on my back and put on my down jacket so that it covers both him and me. My jacket is too small, leaving my middle and upper arms exposed, with only the ends of my arms in my sleeves. The fabric pulls my arms awkwardly backward, and I feel like I'm imitating a T-Rex.

On the walk in we see trails of cottontails and coyotes. The snow is deep, and the cottontail tracks are following in our packed-down rut in the center of the paths from previous visits. At the camera, I swap out the batteries and download the pictures onto my laptop. No bobcat. We head down to the pile of mice in houses we've dumped twenty yards away and add a few more to the pile. There, next to the mice, is a fresh bobcat trail. I remember weeks earlier seeing a bobcat trail following almost exactly the same path into a dense stand of glossy buckthorn, an invasive shrub. I pause for a minute and consider the prominent pin oak standing above the buckthorns. Peeking my head into the thicket, I see that the trunk of the oak is covered in moss.

It's the biggest tree around, near a trail intersection, with a fuzzy surface great for retaining smells; this must be a bobcat scent post. With Alder still on my back, I crawl in to have a closer look. Including the trail we follow in, at least three bobcat trails emanate from the tree. I might be missing a trail, or maybe two of the trails merge into one—because there should be an even number.

I can clearly see where the bobcat paused, squaring up its feet as it sprayed urine backward onto the mossy trunk. I lean my nose in and am rewarded with the pungent smell of cat. Two feet from the tree is a little rectangular hunting lay in the snow where the cat sat hidden in the brush, looking through the edge into the open. Inspecting the individual footprints, my first impression is that there are two different cats—a slightly larger male with rounded toes and a slightly smaller female with more tapered toes.

Now with a fresh set of batteries, the camera is still aimed at the wide trail that forms the center of this chapter's opening image. I have in my mind a picture of the bobcat strolling down the center of the path at that very spot. There was, after all, a bobcat trail in the snow when I took the central photo in this chapter's opening image; it's just so subtle that it doesn't show up in the image. And that spot is, after all, where there used to be a bobcat latrine a decade ago—where dozens of scats were piled in the center of the trail as a blaring message to the world.

But most of the time, the trails of the bobcats I follow in The Orchard, just like the trails of the rabbits, dart in and out of the shrubs. They cross the paths back and forth and sometimes follow the path center for a short time before heading back into the thickets.

Maybe I'm trying to capture a picture of a bobcat in a spot where bobcats don't really want to be. The center of that path is more of a place that coyotes and foxes feel comfortable (figs. 10.10–10.12). Recently, I've been considering placing catnip in the path to lure the cats into the picture. But that feels like cheating. I want to honestly capture nature how it really is. And now there's a natural scent lure attracting bobcats right here on this tree, made by the cats themselves.

Leaving the pin oak, I trudge through the snow back to the camera where it is strapped to a thick bittersweet vine. I unglove my hands and work quickly to free the camera before the cold cripples my fingers. Then I carry the camera back to the pin oak in the thicket fifteen yards away. I set it up on a small tree a few feet from the oak, aimed right at the spot where the urine is sprayed. I can already see the photograph of a bobcat standing in front of this tree marking it, and it makes me smile.

Figure 10.10. A coyote in The Orchard.

I walk back to the spot where the camera used to be and start packing up my computer. Perhaps it's because the sun has shifted in the sky, or perhaps it's because I'm more relaxed, but suddenly a big perfect bobcat footprint pops out at me in the snow a foot from where the camera had just been. The cat had actually stepped right in front of the camera—just like those bastards who stole our bear—I guess it was after the batteries had died.

Why hadn't I noticed any of this trail before? As I look for more tracks, I realize that the cat was mostly walking in the deep footprints of a deer, matching its strides as best it could so as to avoid unnecessarily having to punch new holes in the snow. Oh well, I suppose the scent post is still a better bet for the camera.

Still on my back and ready for a nap, Alder has grown impatient with all my bending, fiddling, and staring at the ground. His fussing spurs me to move on. I finish packing up, hoist my equipment onto my shoulders, and start the trek back through the snow, walking along the trail that separates the aspen forest from the invasive tangle. Step after step through the deep snow, I grow exhausted. I can really feel why the rabbits were using the packed-down rut—even though it's an exposed travel route they would never have chosen otherwise. And I can feel the tired bobcat stepping in the deer tracks, which also led into the packed-down rut. If only it had batteries, the camera aimed at that rut would have captured all the animals. Just as fences funnel salamanders into the tunnel under the road, the snow is funneling all The Orchard's animals into that rut. For predicting animal movements, that snow rut is a temporary goldmine that will soon melt away.

Figure 10.11. A gray fox in The Orchard.

Figure 10.12. A red fox in The Orchard.

Figure 10.13. A bobcat by its scent tree in The Orchard, perhaps looking up at a squirrel.

A few feet from my car, I click the remote to unlock the door. I pause. Alder has fallen asleep on the walk. I see the picture that the tracks painted of the cat marking the pin oak. The scent post will be there long after the snow melts—but not that snow rut in the trail center. I turn from my car and run down to move the camera once more from the scent post back to the spot where it was looking at the rut in the center of the path. I need to talk to my therapists about decision making.

A few days later a warm rain melts all the snow. With the rut now gone, I move the camera back to the scent post. After a month looking at the center of the path, the only predators the camera captured were coyotes and gray foxes. Within a week of looking at the pin oak twenty yards away, the camera captured three bobcat visits (fig. 10.13). Satisfied, I put the camera back on the main path and, after a while, catch a few cats there as well.

GROUNDS

I wonder if the bobcats in the photos I've captured are the off-spring of the bobcats that kept an eye on us in college two decades ago—perhaps the little kittens we tracked here after we graduated. I wonder, too, what's next for these bobcats? How much longer will The Orchard be their home? Will forest succession finally convert the thicket to a forest?

I want the bobcats to stay here forever. But The Orchard that we love—the chaotic tangle—is just a stage in transition. Like racing rocks to a window, there is blinding joy in this moment, but it's bound to end. One way or another, that rock is going out. Does the metaphor work for all of conservation? For all of life?

The immediate threat to this habitat is the fleet of chainsaws and brush hogs cleaning it up. Over the past couple decades, two huge new parking lots have already taken big bites out of The Orchard, and it's been cut in half by a highway-like paved trail lined by grassy shoulders and thirty streetlights. Now they're cutting down the invasive thicket that shelters the bobcats. Is there a specific management plan that the grounds crew is executing with specific conservation goals? I want to know.

One morning after we drop Juno off at school, Alder and I visit the university physical plant, the office that heads up groundskeeping. We walk through the fancy modern entrance, surrounded by solar panel displays and free physical plant schwag. Alder makes friends with Alice at the front desk and another staff member standing nearby, but I'm having a hard time explaining my question. To start, nobody seems to know that The Orchard exists. I'm not talking about new building construction, it's a question about ecological management. You know, landscaping?

Eventually, Alice decides to lead us out back to the "Sustainability Trailer." We walk through a maze of hallways and then outside through a rear metal exit door. Crammed between buildings in a corner of a parking lot is a little white-sided modular office trailer. Once inside, Alice calls out, "Hello?" She checks the main office, then looks through all the cubicles for anybody to talk to.

She turns back to lead us out. "OK, nobody's here right now, but maybe you want to call and leave a message for Ezra."

Just then an ancient man materializes from the last cubicle that Alice had checked. With a slow drawl, he asks, "Hello, can I help you at all?"

"Oh hi! Heh! This gentleman was—is here seeking information about the—what's going up on—what's going on, on, um, Orchard Hill."

"OK. Yeah."

"Can you help him?"

"Uh, I don't have access to that, I'm sorry. Uh, Kyle is in today, he's poking arou—"

"Who?"

"Kyle is the one he'll want to talk to."

"Karla?"

"Kyle, Kyle Lawson."

"Karl Larson?"

"Yeah."

"OK, he's in?"

"Yeah, he's in."

"OK, I haven't seen him. Um, I'll just, I'll try to get him his, do you have a card, one of his cards?"

"He's probably up with, uh—"

"Upstairs?"

"—Jason probably, Jason, maybe, you know him?"

"Yeah, I do. Thank you."

"He's probably talking with him. I'll tell him, uh, if he walks in that you're looking for him, so where would—"

"Excuse me?"

"If he comes in I'll tell him you're looking for him."

"Yeah, I, my name is Alice, I work at the front desk."

"Oh, OK."

"OK?"

"OK."

"Thank you."

Friendly as they are, it's pretty clear to me that I'm not going to get anywhere by trying to engage the bureaucracy in this office complex. As Alice leads me back through other offices in the main building, she laments that folks are in meetings all day long, which is why it's hard to track down people and information. Alice is hopeful that the assistant director of Custodial & Grounds will be able to offer me insights, but she's not in her office. A few days later, when I catch the assistant director on the phone, she doesn't know anything about the operation in The Orchard either.

Before we leave Facilities, back at the front desk, Alice hands us a pile of free plastic cups and six pads of sticky note paper emblazoned with the university insignia and the phrase, "We're Here to Help!" She gives Alder a green lollypop and me some phone numbers and email addresses.

Exiting the building, I have a strong urge to cleanse myself. I feel like I've been coated in a gray layer of dust that's slowing my joints and smothering my skin. It's not a physical dust, it's the emotional debris that saddles me whenever I try to deal with large, Kafkaesque institutions that feel so removed from reality. We drive straight to The Orchard.

On the path to the camera sits a big orange tractor. Inside sits Steve. A big friendly man. He asks if I heard the bird squawking, and then he makes a noise that reminds me of a blue jay. He's been mowing The Orchard for fifteen years. He sees bobcats all the time, all different sizes. Once he saw a big one followed by three little ones cross in front of his tractor. With as much enthusiasm, he tells me he's seen red cardinals in here too. He tells about the deer that know him and his tractor and the hawks that circle around while he mows, waiting to scoop up the mice in his wake.

He tells me of the mountain lion he took pictures of behind his father's house and the mountain lion tracks he saw nearby. Though wildlife officials deny it, Steve's pictures prove that mountain lions live here. This piques my interest, as fifteen years ago I spent over a year trying to verify the presence of mountain lions—also known as cougars, panthers, catamounts, golden ghosts, and a variety of other local names—here in the Northeast.

Eastern cougars are considered officially extinct, having been absent from our landscape since the last of them were intentionally exterminated, with bounties and mass killings, over a hundred years ago. But locals up and down the East Coast still see them every year.

It wasn't hard for me to collect over sixty sightings from folks who had seen mountain lions. Some of these stories were very convincing—biologists and hunters detailing extended close encounters in broad daylight. During that year, I consumed all the literature on the topic, some of which is published in a pseudoscientific journal called *Cryptozoology*, alongside studies of bigfoot, the Loch Ness monster, chupacabras, and other unlikely creatures.

Toward the end of that year, I got to pet a live mountain lion down in West Virginia. It purred loudly and aggressively as it rubbed its cheeks against my fingers through a chain-link fence.

The visit to the captive cat was part of a field trip during the second-ever Eastern Cougar Conference, filled with serious academics, riled-up nutjobs, and lots of in-between folks. One thing unified most attendees of the conference: we wanted to believe, in defiance of the official dogma, that eastern cougars still existed. Then I saw one myself.

I was on my way to Nashville, having just left Cumberland Gap National Historic Park on the border of Kentucky and Tennessee. Like all parks along the Appalachians, it too has had its share of eastern cougar sightings. I reached I-40, which runs the length of Tennessee, at about 11:00 AM and headed west to Nashville. It was a clear, sunny day, and traffic on the interstate was light.

Cruising along at sixty-five miles per hour, I noticed an animal creeping out onto the road from the shoulder, up on a small rise about 150 yards in front of me. At first I thought it was a deer. It was big and had a light tan color. But it wasn't quite a deer. It was lower, and it had a long, swooping tail. As I approached, the form became clearer and clearer. I slowed my car until I was going about twenty-five miles per hour and the animal was about fifteen yards in front of my car. The golden animal, with its muscular front and elegant tail, was creeping along in a slow, steady, catlike walk. "Oh my god, I can't believe it! I'm actually seeing a mountain lion!" Clear as day. It was incredible.

And then, when it was about ten yards in front of me, and my car was almost at a complete stop, suddenly, it was a dog. A dog that just happened to be walking funny. A dog that just happened to be tan. A dog that just happened to have a long, swooping tail. I guess, even in the best of circumstances, the best of observers can make mistakes. My eyes had taken in a few bits of information, and my brain had filled in all the holes in between, painting a perfectly complete mountain lion.

In life we don't really see that much. Mostly our brains are just painting pictures. The ability of people to see and hear things is often limited not by our eyes and our ears but instead by our ability to interpret clues and understand context. As a kid, I used to help my naturalist mentor survey for frogs on the roads while driving along rural highways in the rain at night. Years later, driving my college students in a big van on a rainy nighttime salamander adventure, I spotted a tiny spring peeper on the road ahead, and the students were astounded at my eyesight. But I don't have great eyes. I didn't really see the whole frog. It's just that I've trained my brain to know the difference between how light bounces off wet rocks and wet frogs in the road.

Once, while playing ultimate Frisbee, I had an ocular migraine that robbed my ability to see what was directly in the center of my vision. Instead of seeing a black hole, my brain stitched together the things on either side. When I looked at the face of any player on the field, in the place where their head should have been, their neck just ended. Above the neck I simply saw the grass in the unbroken field behind them. My brain was doing its best with the information it had.

Over the past few decades, there have been a handful of pieces of physical evidence—like scats and hair samples—of mountain lions in the Northeast. But the DNA of some of those animals was traced back to South America—apparently South American mountain lions are more popular in the illicit pet trade, and sometimes people let their pet mountain lions go. One mountain lion was hit by an SUV on a Connecticut highway a few years ago, but that animal had just made a 2,000 mile trip from the Black Hills of South Dakota.

Is there a self-sustaining population of mountain lions here? The habitat seems to be here in the East for them to survive, but it's hard to find any signs of them. When you go out West, it's easy to find definitive tracks and scats of mountain lions. Slowly but surely, western cougars are migrating east, and someday soon there will likely be established cougar populations in the East, if not descendants of the original eastern cougar. But not yet.

Why are we so eager to see mountain lions in the East? Once they are here in big numbers, we'll probably be terrified of them. But for now we want to believe in the mythic power of these gorgeous creatures to overcome all odds and hide in plain sight. Partly, it's connected to why I love tracking. I tend to get more of a thrill from seeing animal footprints than from seeing the animal itself. The tracks have intrigue and mystery and clues to follow and treasures to discover. When I see the animal, it's just, well, an animal. It's vulnerable flesh and blood meekly trying to survive. It peers back at me with disdain. But when I'm still following a trail, the end of it might yet lead me to God herself. If eastern cougars really have persisted, then anything could be possible. Meanwhile there is another real, flesh-and-blood, spectacular wild cat still thriving in our midst, unseen by almost everyone. Can we pin our dreams to the bobcat?

When I ask Tractor Steve to send me a copy of his mountain lion photograph, he demurs. It was a long time ago, the picture would be hard to find, and it wasn't that much to look at anyway. Like most sightings of eastern cougars, this all-but-perfect account ends just shy of being provable.

Steve and I move back to talking about, as he sees it, this little sanctuary that he's lucky enough to have worked in for the past fifteen years. When I ask him about the invasive plant cleanup, he gives me the same sort of response that everyone else has given me. They're cleaning up the invasive species, what else is there to say? It's just something helpful the crew is doing to keep themselves busy in the slow winter months. Isn't it obvious why?

I mean, yes, of course it's obvious. Nobody, especially not university donors and parents, wants to see a messy tangle of shrubs. That's why they're starting on the part visible from the road. And invasive species are bad. This is a way to help the environment. It's a win for everyone. What's to question?

But for me it's not obvious. When parents and donors look at the thicket, they see chaos. I see bobcats. Without the experience of following bobcats day after day, a brain just can't paint the full picture from the few bits of light reflecting off the rose thorns.

Juno says he's going to write a letter to the university to save The Orchard. I share the impulse, but, at the same time, I know it's driven by a misleading sense of nostalgia. It's hard to hold on to the perspective that The Orchard as we see it today, just like every other piece of nature, is, and must be, a fleeting thing. Besides, the invasives are so robust that the effort to remove them will likely fail. Still, I break out some crayons and one of the "We're Here to Help" pads, and we start drafting a letter.

What will be there after the work crews are done? I imagine big trees standing watch over a grassy lawn. Simple, clean, and devoid of wildlife habitat. It's true that the ecosystem now here isn't filled with native plants. But a grassy lawn is no better. These roses are providing habitat for cottontails, skunks, opossums, gray foxes, red foxes, fishers, coyotes, and, my favorite, bobcats. Yes, multiflora rose invades and wrecks many forests, and non-native species in general do a terrible job of supporting native insect diversity. But what was it that they invaded here? Not a native forest.

Standing under the aspens, the *Field Naturalist* students notice patterns on the aerial photographs from fifty years ago (see fig. 10.2). On the west side, the ground is flat and smooth as one would expect a farm field to look. But on the east side, small dark dots, barely perceptible, are formed in neat rows. Today, from the ground, each of these dots looks like a giant Snuffleupagus. Fifty years ago, these were manicured rows of apple trees.

Then the field and the apple trees were abandoned. In the field, wind-dispersed seeds rained down on the dirt. Maple, ash, aspen. As the young seedlings raced skyward, they fought off the voracious voles. Spewing defensive compounds, the aspens got a leg up on the competition and survived to form a young forest.

Over among the rows of apple trees, flocks of robins, cedar waxwings, and bluebirds sat in the apple branches feasting on fruit. Their poop rained down, filled with seeds of privet, rose, bittersweet, grape, and all the other small fruits made for the birds. Back when The Orchard was meticulously pruned and maintained, some of these fruit-bearing shrubs crouched under the apple trees and horse jumps where the lawnmowers couldn't quite reach. When the mowers stopped running, the shrubs lying in wait sprang from every corner, marched across the land, crawled over the aging apple trees, and suppressed any tree seedling that tried to grow.

Thus, it was the teeth of herbivores and mowers from decades ago that shaped today's Orchard where bunnies and bobcats thrive. So it is only fitting that similar teeth now shape the future.

Well, I suppose the teeth are merely the workers out on the ground. Ultimately, these teeth are responding to the choices of human society. What happens to The Orchard in the long run really depends on the whims of the bureaucracy that owns it. Here, as in much of our world, nature's fate is shaped largely by decision makers sitting within a forest of drywall who have never set foot in the actual forests their policies will impact—people who feel no real relationship with nature. But relationships are the heart of humanity. Whether we choose to save the cat, the whale, the salamander, or the mouse depends on which creatures we've formed a relationship with.

* * *

The world is waiting to know: How will we choose to value nature? More fundamentally, how will we perceive nature? Will you see The Orchard as a mess or a wildlife sanctuary? Will you write off the thicket as a worthless web of invasive species or cherish it as home to bobcats? Ought it be tamed or set free? What stories will you make from the places around you? Will you wander off the map and get to know your landscapes? Will you forge connections with wild places? Will you bang rocks together to see how they break? Will you ask the oaks and pines for a story? Will you taste pine needle tea? Will you listen as the birds reflect on you? Will you wonder why the scat is placed where it is? Will you forge personal relationships with individual animals, species, and ecosystems? Will you build your life's meaning and fill your memories with the natural world? Will you carry your children on your back up mountains and down rivers, and will you set them free to roam in your nearest wild space so that they grow up as native citizens of nature? I hope so. The way to begin is to set down this book, set down all books, set down all things, and wander, unencumbered, without direction, outside.

MAJOR LESSONS FOR INTERPRETING A LANDSCAPE

- Where did the species at your site originate?
- What personal values do you carry when you evaluate the worth of the things before you?
- What signs of animals do you see, and how do the animals see this landscape?
- Stop and listen to the birds. Learn to use them as interpretive guides.

Author's Note

Caleb, Maya, Alice, and Tractor Steve are real people whose names have been changed to protect their privacy. Bark Hollow and Maggie's Forest are names I created for very particular spots that otherwise don't have names at such a small scale.

Acknowledgments

I should start by thanking the natural world—the ground, the plants, the crawling critters, the flying critters, the winds and weather, the sun and moon, night and day, the unknown mysteries, and all else.

As for the people, I owe a ton of thanks to all the folks who were essential to this project. David Foster at Harvard Forest and Jean Thomson Black at Yale University Press were instrumental in supporting me throughout. Harvard University's Charles Bullard Fellowship in Forest Research and Yale University Press, both the institutions and the people behind the scenes, were of critical importance. Lydia Rogers, Burt Adelman, ASLE, the Helen Clay Frick Foundation, and an anonymous donor gave generous and crucial support.

Alicia Daniel and the University of Vermont Field Naturalist Program helped me create my version of their course—the model for this book. Without this seed the book would never have grown.

I owe much to the authors whose books inspired me, especially Tom Wessels.

Also a big thanks to Charley Eiseman, Julia Blyth, Susannah Lerman, Tekla Harms, David Foster, Tamar Charney, Sharon Charney, Mike Jones, Peter Burn, Aaron Ellison, Paige Warren, Robert Sanford, Kathleen Deselle, Gretchen Otto, Mary Pasti, and an anonymous reviewer for reading and editing parts of the manuscript.

To all the people—friends, colleagues, students, strangers,

landowners, and others—whose stories, work, and land I've borrowed, I owe an enormous debt.

In addition to all the people depicted, there are many more who helped me out: Neil Pederson, Dave Stahle, Will Blozan, and Fred Paillet steered me toward chestnuts. Don Wise, Isaac Larsen, Stephen Mabee, and Selby Cull-Hearth offered geology insights. Chris Davis, Neil Kapitulik, Kate Froburg, and Sara Wisner helped with tiger beetles. Steve Gephard and Theodore Castro-Santos helped with historical fish numbers. Peter Burn helped me understand salt marshes. Tony D'Amato guided me on hemlocks and old growth. Dave Orwig and Annie Paradis gave me the skinny on adelgids. Kathryn Leigh tracked down historical maps. Larry Hunter helped troubleshoot physics concepts, Tom French with vole ID, Rodent Twitter with mice ID. Thanks to Walker Korby's basement freezer, Chelse Leven's Antarctic reports, all the Woodsies, and everyone else I've missed.

Thanks so much to all my teachers, including tracker Sue Morse for her mnemonics (X marks the Spot, catitude, the circle-three-tracks trick, etc.), John McCarter, Mark Elbroch, Larry Winship, Louis Levine, and many, many more.

Shout-outs to the Jones Library, Sunderland Historic Society, Sunderland Library, Chester Springs Library, Henrietta Hankin Library, Five-College Libraries, Bryn Mawr Library, and Kimberton Whole Foods for places to work and develop ideas, not to mention the brilliance of interlibrary loan. Shout-outs too to all the private and public outdoor spaces I visited, including a long list of local, regional, national, and international conservation areas.

*   *   *

A special thanks to my family for constant support. Mom, dad, brother, sister. Sydne for many ideas and directions.

And to my guides, cocreators, and first audience, with enormous love: Alder and Juno.

Bibliography

GENERAL REFERENCE GUIDES (USED THROUGHOUT)
Field Guides

Eiseman C. 2018. *Leafminers of North America*. Privately published e-book. Available at: http://charleyeiseman.com/leafminers/.

Eiseman C, Charney ND & Carlson J. 2010. *Tracks & Sign of Insects & Other Invertebrates: A Guide to North American Species*. Stackpole Books.

Elbroch M. 2003. *Mammal Tracks & Sign: A Guide to North American Species*. Stackpole Books.

Elbroch M & Marks E. 2001. *Bird Tracks & Sign: A Guide to North American Species*. Stackpole Books.

Elbroch M & Rinehart K. 2011. *Behavior of North American Mammals*. Houghton Mifflin Harcourt.

Elias TS & Dykeman PA. 1982. *Field Guide to North American Edible Wild Plants*. Outdoor Life Books.

Maine Forest Service. 2008. *The Forest Trees of Maine, Centennial Edition 1908–2008*. Maine Department of Conservation.

Newcomb L. 1977. *Newcomb's Wildflower Guide*. Little, Brown.

Peterson L. 1977. *A Field Guide to Edible Wild Plants of Eastern and Central North America*. Peterson Field Guide Series. Houghton Mifflin Harcourt.

Petrides GA & Wehr J. 1988. *A Field Guide to Eastern Trees: Eastern United States and Canada. Peterson Field Guide Series.* Houghton Mifflin.

Rezendes P. 1999. *Tracking & the Art of Seeing: How to Read Animal Tracks & Sign.* Harper Perennial.

Roberts DC. 1996. *A Field Guide to Geology: Eastern North America. Peterson Field Guide Series.* Houghton Mifflin Company.

Digital Platforms

Integrated Taxonomic Information System (ITIS). http://www.itis.gov.

Native Plant Trust: Go Botany. https://gobotany.nativeplanttrust.org/.

Schweitzer PN. 2011. Combined geologic map data for the conterminous US derived from the USGS state geologic map compilation. http://mrdata.usgs.gov/geology/state/geol_poly.zip.

Other Texts

Barbour MG, Burk JH, et al. 1999. *Terrestrial Plant Ecology.* Addison Wesley Longman.

Burns RM & Honkala BH. 1990. *Silvics of North America. Forest Service Agricultural Handbook 654.* United States Department of Agriculture.

Molles MC. 2010. *Ecology: Concepts and Applications. Fifth Edition.* McGraw-Hill Higher Education.

Plummer CC, Carlson DH & McGeary D. 2007. *Physical Geology. Eleventh Edition.* McGraw-Hill Higher Education.

Swain PC & Kearsley JB. 2000. *Classification of the Natural Com-*

munities of Massachusetts. Natural Heritage & Endangered Species Program, Massachusetts Division of Fisheries and Wildlife.

Wessels T. 1997. *Reading the Forested Landscape. A Natural History of New England*. Countryman Press.

1. HOME
Trees as Insulators

Haesen S, Lembrechts JJ, et al. 2021. ForestTemp – Sub-canopy microclimate temperatures of European forests. *Global Change Biology*, 27(23).

Heisler GM. 1986. Energy savings with trees. *Journal of Arboriculture*, 12(5).

Urban Ecology

Aronson MF, Lepczyk CA, et al. 2017. Biodiversity in the city: Key challenges for urban green space management. *Frontiers in Ecology and the Environment*, 15(4).

Badyaev AV, Young RL, et al. 2008. Evolution on a local scale: Developmental, functional, and genetic bases of divergence in bill form and associated changes in song structure between adjacent habitats. *Evolution*, 62(8).

Blair RB. 1996. Land use and avian species diversity along an urban gradient. *Ecological Applications*, 6(2).

Fischer JD & Miller JR. 2015. Direct and indirect effects of anthropogenic bird food on population dynamics of a songbird. *Acta Oecologica*, 69.

Kane B, Warren PS & Lerman SB. 2015. A broad scale analysis of tree risk, mitigation and potential habitat for cavity-nesting birds. *Urban Forestry & Urban Greening*, 14(4).

Lerman SB, Contosta AR, et al. 2018. To mow or to mow less: Lawn mowing frequency affects bee abundance and diversity in suburban yards. *Biological Conservation*, 221.

Lerman SB & Warren PS. 2011. The conservation value of residential yards: Linking birds and people. *Ecological Applications*, 21(4).

McKinney ML. 2002. Urbanization, biodiversity, and conservation. *BioScience*, 52(10).

Pickett ST, Cadenasso ML, et al. 2001. Urban ecological systems: Linking terrestrial ecological, physical, and socioeconomic components of metropolitan areas. *Annual Review of Ecology and Systematics*, 32(1).

Shochat E, Lerman SB, et al. 2010. Invasion, competition, and biodiversity loss in urban ecosystems. *BioScience*, 60(3).

US Census Bureau. 2010. *2010 Census Urban and Rural Classification and Urban Area Criteria*. US Department of Commerce, Washington, DC. http://www.census.gov.

Arizona Mud Turtles

American Turtle Observatory. 2016. *2016 Year-End Report*. http://americanturtles.org.

Ecological Footprints

Bailis R, Drigo R, et al. 2015. The carbon footprint of traditional woodfuels. *Nature Climate Change*, 5(3).

Barto D, Cziraky J, et al. 2009. An integrated analysis of the use of woodstoves to supplement fossil fuel–fired domestic heating. *Journal of Natural Resources & Life Sciences Education*, 38.

Eaton RL, Hammond GP & Laurie J. 2007. Footprints on the

landscape: An environmental appraisal of urban and rural living in the developed world. *Landscape and Urban Planning*, 83(1).

Holden, E. 2004. Ecological footprints and sustainable urban form. *Journal of Housing and the Built Environment*, 19(1).

Jones C & Kammen DM. 2014. Spatial distribution of US household carbon footprints reveals suburbanization undermines greenhouse gas benefits of urban population density. *Environmental Science & Technology*, 48(2).

Wilson J, Spinney J, et al. 2013. Blame the exurbs, not the suburbs: Exploring the distribution of greenhouse gas emissions within a city region. *Energy Policy*, 62.

Boston Drumlins

Crosby IB. 1934. Evidence from drumlins concerning the glacial history of Boston Basin. *Bulletin of the Geological Society of America*, 45(1).

Field Naturalist Teaching—The UVM Model

Hagenbuch BE. 2006. *Reconceptualizing Natural History Study in Higher Education: Perspectives from the Field*. PhD dissertation, Environmental Studies, Antioch New England Graduate School.

Giving Thanks

Michaelson J. 2007. *God in your body: Kabbalah, mindfulness and embodied spiritual practice*. Jewish Lights.

Stokes J, Kanawahienton, et al. 1993. *Thanksgiving Address: Greetings to the Natural World*. English Version. Six Nations Indian

Museum and the Tracking Project. https://americanindian.si.edu
/environment/pdf/01_02_Thanksgiving_Address.pdf.

Swamp J. 1997. *Giving Thanks: A Native American Good Morning Message*. Lee & Low Books.

2. LAND
Orientation, Aspect, and Evaporative Demand

Charney ND, Babst F, et al. 2016. Observed forest sensitivity to climate implies large changes in 21st century North American forest growth. *Ecology Letters*, 19(9).

Potzger JE. 1939. Microclimate and a notable case of its influence on a ridge in central Indiana. *Ecology*, 20(1).

Night Sky

Chartrand MR. 1991. *National Audubon Society Field Guide to the Night Sky*. Alfred A. Knopf.

Population Viability Analyses

Fiedler PL. 1985. Heavy metal accumulation and the nature of edaphic endemism in the genus *Calochortus* (Liliaceae). *American Journal of Botany*, 72(11).

Fiedler PL. 1987. Life history and population dynamics of rare and common mariposa lilies (*Calochortus* Pursh: Liliaceae). *Journal of Ecology*, 75(4).

Gawler SC, Waller DM & Menges ES. 1987. Environmental factors affecting establishment and growth of *Pedicularis furbishiae*, a rare endemic of the St. John River Valley, Maine. *Bulletin of the Torrey Botanical Club*, 114(3).

American Chestnuts

Anagnostakis SL. 1987. Chestnut blight: The classical problem of an introduced pathogen. *Mycologia*, 79(1).

Faison EK & Foster DR. 2014. Did American chestnut really dominate the eastern forest? *Arnoldia*, 72(2).

Hoadley RB. 1990. Identifying wood: Accurate results with simple tools. Taunton Press.

Lorimer CG. 1980. Age structure and disturbance history of a Southern Appalachian virgin forest. *Ecology*, 61(5).

Milgroom MG & Cortesi P. 2004. Biological control of chestnut blight with hypovirulence: A critical analysis. *Annual Review of Phytopathology*, 42.

National Park Service. 2007. *Great Smoky Mountains National Park Briefing Statement, April 2007. Summary of Forest Insect and Disease Impacts.* http://www.nps.gov.

Oosting HJ & Bourdeau PF. 1955. Virgin hemlock forest segregates in the Joyce Kilmer Memorial Forest of western North Carolina. *Botanical Gazette*, 116(4).

Paillet FL. 1988. Character and distribution of American chestnut sprouts in southern New England woodlands. *Bulletin of the Torrey Botanical Club*, 115(1).

Weaver GR. 2003. Chestnut ghosts: Remnants of the primeval American chestnut forest of the Southern Appalachians. *Chestnut*, XVI(2).

Survival

Brown T & Morgan B. 1983. *Tom Brown's Field Guide to Wilderness Survival*. Berkley Books.

Piantadosi CA. 2003. *The Biology of Human Survival: Life and Death in Extreme Environments*. Oxford University Press.

Sullivan RJ. 1993. Accepting death without artificial nutrition or hydration. *Journal of General Internal Medicine*, 8(4).

Bedrock Geology

Benson RN. 1992. Map of Exposed and Buried Early Mesozoic Rift Basins/Synrift Rocks of the U.S. Middle Atlantic Continental Margin (map): Delaware Geological Survey Miscellaneous Map Series 5, scale 1:1,000,000.

Little RD. 1986. *Dinosaurs, Dunes, and Drifting Continents: The Geohistory of the Connecticut Valley*. Valley Geology Publications.

Hepatica in Medicine

Foster S & Duke JA. 2000. *A Field Guide to Medicinal Plants and Herbs of Eastern and Central North America*. Peterson Field Guide Series. Houghton Mifflin Harcourt.

Soils and Nutrients

Burkhart EP. 2013. American ginseng (*Panax quinquefolius* L.) floristic associations in Pennsylvania: Guidance for identifying calcium-rich forest farming sites. *Agroforestry Systems*, 87(5).

Conant RT, Easter M, et al. 2007. Impacts of periodic tillage on soil C stocks: A synthesis. *Soil and Tillage Research*, 95(1-2).

Hepler PK. 2005. Calcium: A central regulator of plant growth and development. *The Plant Cell*, 17(8).

Huggett RJ. 1998. Soil chronosequences, soil development, and soil evolution: A critical review. *Catena*, 32(3-4).

Ismail I, Blevins RL & Frye WW. 1994. Long-term no-tillage effects on soil properties and continuous corn yields. *Soil Science Society of America Journal*, 58(1).

Record S, Kobe RK, et al. 2016. Seedling survival responses to conspecific density, soil nutrients, and irradiance vary with age in a tropical forest. *Ecology*, 97(9).

Searcy KB, Wilson BF & Fownes JH. 2003. Influence of bedrock and aspect on soils and plant distribution in the Holyoke Range, Massachusetts. *Journal of the Torrey Botanical Society*, 130(3).

Shaul O. 2002. Magnesium transport and function in plants: The tip of the iceberg. *Biometals*, 15(3).

3. WATER

White Pine—Tree of Peace

Fenton WN. 1998. *The Great Law and the Longhouse: A Political History of the Iroquois Confederacy* (Vol. 223). University of Oklahoma Press.

Maulucci MSR. 2010. Invoking the sacred: Reflections on the implications of ecojustice for science education. In Tippins DJ, Mueller MP, et al., eds., *Cultural Studies and Environmentalism* (pp. 43–49). Springer, Dordrecht.

Pitch Pine

Ledig FT & Fryer JH. 1972. A pocket of variability in *Pinus rigida*. *Evolution*, 26(2).

Ledig FT, Hom JL & Smouse PE. 2013. The evolution of the New Jersey pine plains. *American Journal of Botany*, 100(4).

Pausas JG. 2015. Evolutionary fire ecology: Lessons learned from pines. *Trends in Plant Science*, 20(5).

Clements FE. 1916. *Plant Succession: An Analysis of the Development of Vegetation*. No 242. Carnegie Institution of Washington.

Collin A, Messier C, et al. 2017. Low light availability associated with American beech is the main factor for reduced sugar maple seedling survival and growth rates in a hardwood forest of southern Quebec. *Forests*, 8(11).

Davis MB. 1981. Quaternary history and the stability of forest communities. In West DC, Shugart HH & Botkin DB, eds., *Forest Succession: Concepts and Application* (pp. 132–153). Springer.

Gleason HA. 1926. The individualistic concept of the plant association. *Bulletin of the Torrey Botanical Club*, 53(1).

Hane EN, Hamburg SP, et al. 2003. Phytotoxicity of American beech leaf leachate to sugar maple seedlings in a greenhouse experiment. *Canadian Journal of Forest Research*, 33(5).

Lichter J. 2000. Colonization constraints during primary succession on coastal Lake Michigan sand dunes. *Journal of Ecology*, 88(5).

Olson JS. 1958. Lake Michigan dune development 2. Plants as agents and tools in geomorphology. *The Journal of Geology*, 66(4).

Peattie DC. 1930. *Flora of the Indiana Dunes*. Field Museum of Natural History.

Takahashi K, Arii K & Lechowicz MJ. 2010. Codominance of Acer saccharum and Fagus grandifolia: The role of Fagus root sprouts along a slope gradient in an old-growth forest. *Journal of Plant Research*, 123(5).

Thoreau HD & Emerson RW. 1887. *The Succession of Forest Trees and Wild Apples*. Houghton, Mifflin.

Williams JW, Shuman BN, et al. 2004. Late-Quaternary vegetation dynamics in North America: Scaling from taxa to biomes. *Ecological Monographs*, 74(2).

Anderson RC. 2006. Evolution and origin of the Central Grassland of North America: Climate, fire, and mammalian grazers. *Journal of the Torrey Botanical Society*, 133(4).

Askins RA. 1999. History of grassland birds in eastern North America. *Studies in Avian Biology*, 19.

Channell R & Lomolino MV. 2000. Dynamic biogeography and conservation of endangered species. *Nature*, 403.

Commonwealth of Massachusetts. n.d. *An action plan for the conservation of state-listed obligate grassland birds in Massachusetts.* Accessed January 17, 2019, http://www.mass.gov.

Donlan CJ. 2005. Re-wilding North America. *Nature*, 436.

Donlan CJ, Berger J, et al. 2006. Pleistocene rewilding: An optimistic agenda for twenty-first century conservation. *The American Naturalist*, 168(5).

Faison EK, Foster DR, et al. 2006. Early Holocene openlands in southern New England. *Ecology*, 87(10).

Foster DR & Motzkin G. 2003. Interpreting and conserving the openland habitats of coastal New England: Insights from landscape history. *Forest Ecology and Management*, 185(1-2).

Lesica P & Allendorf FW. 1995. When are peripheral populations valuable for conservation? *Conservation Biology*, 9(4).

Norment C. 2002. On grassland bird conservation in the Northeast. *The Auk*, 119(1).

Vickery PD & Dunwiddle PW, eds. 1997. *Grasslands of Northeastern North America: Ecology and Conservation of Native and Agricultural Landscapes.* Massachusetts Audubon Society.

Wells JV, Robertson B, et al. 2010. Global versus local conservation focus of US state agency endangered bird species lists. *Plos One*, 5(1).

Fire, Pine Barrens, and Land-Use History

Clark KH & Patterson III WA. 2003. *Fire Management Plan for Montague Plain Wildlife Management Area*. Department of Natural Resources Conservation, University of Massachusetts Amherst.

Compton JE, Boone RD, et al. 1998. Soil carbon and nitrogen in a pine-oak sand plain in central Massachusetts: Role of vegetation and land-use history. *Oecologia*, 116(4).

Motzkin G, Foster DR, et al. 1996. Controlling site to evaluate history: Vegetation patterns of a New England sand plain. *Ecological Monographs*, 66(3).

Motzkin G, Patterson III WA & Foster DR. 1999. A historical perspective on pitch pine–scrub oak communities in the Connecticut Valley of Massachusetts. *Ecosystems*, 2(3).

Pauley TK. 2008. The Appalachian inferno: Historical causes for the disjunct distribution of *Plethodon nettingi* (Cheat Mountain Salamander). *Northeastern Naturalist*, 15(4).

Pyne SJ. 2019. *The Northeast: A Fire Survey*. University of Arizona Press.

The Glacial Lake

Bruchac M. (2005). The Geology and Cultural History of the Beaver Hill Story. *Raid on Deerfield: The Many Stories of 1704*. Available at: http://repository.upenn.edu/anthro_papers/144.

Goebel T, Waters MR & O'Rourke DH. 2008. The late Pleistocene dispersal of modern humans in the Americas. *Science*, 319.

Grayson DK. 1991. Late Pleistocene mammalian extinctions in North America: Taxonomy, chronology, and explanations. *Journal of World Prehistory*, 5(3).

Hooke RL & Ridge JC. 2016. Glacial lake deltas in New England record continuous, not delayed, postglacial rebound. *Quaternary Research*, 85(3).

Ridge JC, Balco G, et al. 2012. The New North American Varve Chronology: A precise record of southeastern Laurentide Ice Sheet deglaciation and climate, 18.2–12.5 kyr BP, and correlations with Greenland ice core records. *American Journal of Science*, 312(7).

Stone JR & Ashley GM. 1992. Ice-wedge casts, pingo scars, and the drainage of glacial Lake Hitchcock. Trip A-7 in Robinson, P & Brady JB, eds., *Guidebook for Field Trips in the Connecticut Valley Region of Massachusetts and Adjacent States, Vol. 2.* New England Intercollegiate Geological Conference 84th Annual Meeting, Amherst, Mass., Oct. 9–11, 1992. University of Massachusetts, Geology and Geography Contribution 66(2).

4. CONTEXT

Maps

Evans RT & Frye HM. 2009. *History of the Topographic Branch (Division).* Circular 1341. US Geological Survey.

Hack JT. 1948. Photo-Interpretation in Military Geology. *Photogrammetric Engineering*, 14(4).

Moore L. 2011, May 16. US Topo—A New National Map Series. *Directions Magazine.* Accessed February 28, 2019, https://www.directionsmag.com.

US Department of the Army, Headquarters. 2005. *Map Reading and Land Navigation.* Field Manual 3-25.26. Washington, DC.

Vernal Pools

Wright AH. 1914. North American Anura: Life-histories of the Anura of Ithaca, New York (No. 197). Carnegie Institution of Washington.

Beech Bark Disease

Cale JA, Garrison-Johnston MT, et al. 2017. Beech bark disease in North America: Over a century of research revisited. *Forest Ecology and Management*, 394.

Cale JA & McNulty SA. 2018. Not dead yet: Beech trees can survive nearly three decades in the aftermath phase of a deadly forest disease complex. *Forest Ecology and Management*, 409.

Tupelo

Burckhalter RE. 1992. The genus *Nyssa* (Cornaceae) in North America: A revision. *SIDA, Contributions to Botany*, 15(2).

Keeley JE. 1979. Population differentiation along a flood frequency gradient: Physiological adaptations to flooding in *Nyssa sylvatica*. *Ecological Monographs*, 49(1).

Wen J & Stuessy TF. 1993. The phylogeny and biogeography of *Nyssa* (Cornaceae). *Systematic Botany*, 18(1).

Zhou W, Ji X, et al. 2018. Resolving relationships and phylogeographic history of the *Nyssa sylvatica* complex using data from RAD-seq and species distribution modeling. *Molecular Phylogenetics and Evolution*, 126.

Tree Recruitment

Collins SL. 1990. Habitat relationships and survivorship of tree seedlings in hemlock-hardwood forest. *Canadian Journal of Botany*, 68(4).

Collins SL & Good RE. 1987. The seedling regeneration niche: Habitat structure of tree seedlings in an oak-pine forest. *Oikos*, 48(1).

George LO & Bazzaz FA. 1999. The fern understory as an ecological filter: Emergence and establishment of canopy-tree seed-lings. *Ecology*, 80(3).

Niinemets Ü. 2010. Responses of forest trees to single and mul-tiple environmental stresses from seedlings to mature plants: Past stress history, stress interactions, tolerance and acclimation. *Forest Ecology and Management*, 260(10).

Wetlands

Crafts JM. 1899. *History of the Town of Whately, Mass: Including a Narrative of Leading Events from the First Planting of Hatfield: 1661–1899.* D. L. Crandall.

Millennium Ecosystem Assessment. 2005. *Ecosystems and Human Well-Being: Wetlands and Water, Synthesis.* World Resources Institute.

Mitsch WJ. 2005. Applying science to conservation and res-toration of the world's wetlands. *Water Science and Technology*, 51(8).

Zedler JB & Kercher S. 2005. Wetland resources: Status, trends, ecosystem services, and restorability. *Annual Review of Environment and Resources*, 30.

5. CHANGE
Beavers

Anderson R. 2002. "*Castor canadensis*" (Online), Animal Diver-sity Web. Accessed February 27, 2019, https://animaldiversity.org/accounts/Castor_canadensis/.

Wright JP, Jones CG & Flecker AS. 2002. An ecosystem engi-neer, the beaver, increases species richness at the landscape scale. *Oecologia*, 132(1).

Ice Jams

Beltaos S, ed. 1995. *River Ice Jams*. Water Resources.

Martinson C. 1980. *Sediment Displacement in the Ottauquechee River, 1975–1978*. Special Report No. CRREL-SR-80-20. United States Army Corps of Engineers Cold Regions Research and Engineering Laboratory, Hanover, NH.

Scrimgeour GJ, Prowse TD, et al. 1994. Ecological effects of river ice break-up: A review and perspective. *Freshwater Biology*, 32(2).

River Terraces

Born SM & Ritter DF. 1970. Modern terrace development near Pyramid Lake, Nevada, and its geologic implications. *Geological Society of America Bulletin*, 81(4).

Bull WB. 1990. Stream-terrace genesis: Implications for soil development. *Geomorphology*, 3.

Davis WM. 1909. *Geographical Essays*. Ginn.

Limaye AB & Lamb MP. 2016. Numerical model predictions of autogenic fluvial terraces and comparison to climate change expectations. *Journal of Geophysical Research: Earth Surface*, 121(3).

Mizutani T. 1998. Laboratory experiment and digital simulation of multiple fill-cut terrace formation. *Geomorphology*, 24(4).

Forest Fragmentation, Invasive Species

Andrén H & Anglestam P. 1988. Elevated predation rates as an edge effect in habitat islands: Experimental evidence. *Ecology*, 69(2).

Crooks KR & Soulé ME. 1999. Mesopredator release and avifaunal extinctions in a fragmented system. *Nature*, 400.

Deng WH & Gao W. 2005. Edge effects on nesting success of cavity-nesting birds in fragmented forests. *Biological Conservation*, 126(3).

Forman RT & Deblinger RD. 2000. The ecological road-effect zone of a Massachusetts (USA) suburban highway. *Conservation Biology*, 14(1).

Gillies S, Clements DR & Grenz J. 2016. Knotweed (*Fallopia* spp.) invasion of North America utilizes hybridization, epigenetics, seed dispersal (unexpectedly), and an arsenal of physiological tactics. *Invasive Plant Science and Management*, 9(1).

Magura T, Tóthmérész B & Molnár T. 2001. Forest edge and diversity: Carabids along forest-grassland transects. *Biodiversity & Conservation*, 10(2).

Murcia C. 1995. Edge effects in fragmented forests: Implications for conservation. *Trends in Ecology & Evolution*, 10(2).

Paton PW. 1994. The effect of edge on avian nest success: How strong is the evidence? *Conservation Biology*, 8(1).

Sanders L. 2014. *The Natural History of the Rainbow Beach Conservation Area.* Report for the Northampton's Community Preservation Fund, Northampton, MA.

Endangered Species

Condron A, DeConto R, et al. 2005. Multidecadal North Atlantic climate variability and its effect on North American salmon abundance. *Geophysical Research Letters*, 32(23).

Daley B. 2012, August 5. US bid to return salmon to Connecticut River ends. *Boston Globe*.

Duran DP, Herrmann DP, et al. 2018. Cryptic diversity in the North American *Dromochorus* tiger beetles (Coleoptera: Carabidae: Cicindelinae): A congruence-based method for species discovery. *Zoological Journal of the Linnean Society*, 186(1).

Gephard S & J McMenemy. 2004. An overview of the program to restore Atlantic salmon and other diadromous fishes to the Connecticut River with notes on the current status of these species in the river. *American Fisheries Society Monograph*, 9.

Jones MT. 2009. *Spatial ecology, population structure, and conservation of the wood turtle,* Glyptemys insculpta, *in central New England*. Doctoral dissertation, University of Massachusetts, Open Access Dissertations, 39.

Juanes F, Gephard S & Beland KF. 2004. Long-term changes in migration timing of adult Atlantic salmon (*Salmo salar*) at the southern edge of the species distribution. *Canadian Journal of Fisheries and Aquatic Sciences*, 61(12).

Parrish DL, Behnke RJ, et al. 1998. Why aren't there more Atlantic salmon (*Salmo salar*)? *Canadian Journal of Fisheries and Aquatic Sciences*, 55(S1).

US Fish and Wildlife Service. 1993. *Puritan Tiger Beetle* (Cicindela puritana G. Horn) *Recovery Plan*. Hadley, MA.

US Fish and Wildlife Service. 1998. *Oregon Chub* (Oregonichthys crameri) *Recovery Plan*. Portland, OR.

US Fish and Wildlife Service. 2013. *Draft Post-delisting Monitoring Plan for the Oregon Chub* (Oregonichthys crameri). Portland, OR.

Florida Calusa Mounds

Jones MT, Willey LL & Charney ND. 2016. Box turtles (*Terrapene carolina bauri*) on ancient, anthropogenic shell work islands in the Ten Thousand Islands Mangrove Estuary, Florida, USA. *Journal of Herpetology*, 50(1).

Schwadron M. 2010. *Landscapes of maritime complexity: Prehistoric shell work sites of the Ten Thousand Islands, Florida*. Doctoral dissertation, University of Leicester.

Philosophy of Change

Hesse H. 1998. *Siddhartha*. Translated by Appelbaum S. Courier Corporation.

Kahn CH, ed. 1981. *The Art and Thought of Heraclitus: A New Arrangement and Translation of the Fragments with Literary and Philosophical Commentary*. Cambridge University Press.

Rāhula W. 1974. *What the Buddha Taught*. Grove Press.

6. CHEMICALS
Bedrock Geology and Plate Tectonics

Cooper MP & Mylroie JE. 2015. Geologic and geomorphic history of New York and New England. In *Glaciation and Speleogenesis* (pp. 19–32). Springer, Cham.

Rice WN & Gregory HE. 1908. *Manual of the Geology of Connecticut* (No. 6). Hartford Press, Case, Lockwood & Brainard.

Rodgers J. 1985. Bedrock Geological Map of Connecticut. Connecticut Geological and Natural History Survey.

Skehan JW. 2001. *Roadside Geology of Massachusetts*. Mountain Press.

Tait J, Schätz M, et al. 2000. Palaeomagnetism and Palaeozoic palaeogeography of Gondwana and European terranes. *Geological Society, London, Special Publications*, 179(1).

Zen EA, Goldsmith R, et al. 1983. Bedrock Geologic Map of Massachusetts. US Geological Survey.

Biogeography

Humphries CJ. 1983. Biogeographical explanations and the southern beeches. In Sims RW, Price JH & Whalley PES, eds., *Evo-*

lution, Time, and Space: The Emergence of the Biosphere (pp. 335–365). Academic Press.

Larson A, Wake D & Devitt T. 2007. Salamandridae. Newts and "true salamanders." Version 24, January 2007. The Tree of Life Web Project. http://tolweb.org/.

Matthews JV. 1980. Tertiary land bridges and their climate: Backdrop for development of the present Canadian insect fauna. *The Canadian Entomologist*, 112(11).

Milner AR. 1983. The biogeography of salamanders in the Mesozoic and early Caenozoic: A cladistic-vicariance model. In Sims RW, Price JH & Whalley PES, eds., *Evolution, Time, and Space: The Emergence of the Biosphere* (pp. 431–468). Academic Press.

Pearson MR. 2016. *Phylogeny and systematic history of early salamanders*. Doctoral dissertation, University College London.

Pough FH, Andrews RM, et al. 1998. *Herpetology*. Pearson Prentice Hall.

Tiffney BH. 1985. The Eocene North Atlantic land bridge: Its importance in Tertiary and modern phytogeography of the Northern Hemisphere. *Journal of the Arnold Arboretum*, 66(2).

Torsvik TH, Carlos D, et al. 2002. Global reconstructions and North Atlantic palaeogeography 400 Ma to Recent. In Eide EA, ed., *BATLAS—Mid Norway Plate Reconstructions Atlas with Global and Atlantic Perspectives* (pp. 18–39). Geological Survey of Norway.

Torsvik TH & Cocks LRM. 2013. Gondwana from top to base in space and time. *Gondwana Research*, 24(3-4).

Whiteside JH, Grogan DS, et al. 2011. Climatically driven biogeographic provinces of Late Triassic tropical Pangea. *Proceedings of the National Academy of Sciences*, 108(22).

Zhang P & Wake DB. 2009. Higher-level salamander relationships and divergence dates inferred from complete mitochondrial genomes. *Molecular Phylogenetics and Evolution*, 53(2).

Randomness

Turner MG, Gardner RH & O'Neill RV. 2001. *Landscape Ecology in Theory and Practice.* Springer.

Gull Feeding

Ingolfsson A & Estrella BT. 1978. The development of shell-cracking behavior in herring gulls. *The Auk*, 95(3).

DDT, Malaria, and Osprey

Bierregaard Jr. RO, David AB, et al. 2014. Post-DDT recovery of osprey (*Pandion haliaetus*) populations in Southern New England and Long Island, New York, 1970–2013. *Journal of Raptor Research*, 48(4).

Carson R. 2002. *Silent Spring.* Houghton Mifflin Harcourt.

Hellou J, Lebeuf M & Rudi M. 2012. Review on DDT and metabolites in birds and mammals of aquatic ecosystems. *Environmental Reviews*, 21(1).

Hoffman DJ, Rattner BA, et al., eds. 2002. *Handbook of Ecotoxicology.* Lewis.

Woodside C. 2002, February 3. Where bird guides were born, soon a refuge? *New York Times.*

Tides

Cartwright DE. 2000. *Tides: A Scientific History.* Cambridge University Press.

de Kleer K & de Pater I. 2016. Time variability of Io's volcanic activity from near-IR adaptive optics observations on 100 nights in 2013–2015. *Icarus*, 280.

Desplanque C & Mossman DJ. 2001. Bay of Fundy tides. *Geoscience Canada*, 28(1).

Einstein A. 1920. *Relativity: The Special and the General Theory, a Popular Exposition*. Translated by Lawson RW. Methuen.

Komossa S. 2015. Tidal disruption of stars by supermassive black holes: Status of observations. *Journal of High Energy Astrophysics*, 7.

Ohanian HC. 1989. *Physics. Second Edition*. W. W. Norton.

Panning M, Lekic V, et al. 2006. Long-period seismology on Europa: 2. Predicted seismic response. *Journal of Geophysical Research*, 111.

Wigwams

Chamberlain AF. 1902. Algonkian Words in American English. *Journal of American Folklore*, 15(59).

Nabokov P & Easton R. 1989. *Native American Architecture*. Oxford University Press.

Smell and Pheromones (Spongy Moths)

Acharya L & McNeil JN. 1998. Predation risk and mating behavior: The responses of moths to bat-like ultrasound. *Behavioral Ecology*, 9(6).

Callahan PS. 1975. Insect antennae with special reference to the mechanism of scent detection and the evolution of the sensilla. *International Journal of Insect Morphology and Embryology*, 4(5).

Cameron EA. 1981. On the persistence of disparlure in the human body. *Journal of Chemical Ecology*, 7(2).

*Invasive Genotypes (*Phragmites *and Tiger Salamanders)*

Fitzpatrick BM, Johnson JR, et al. 2010. Rapid spread of invasive genes into a threatened native species. *Proceedings of the National Academy of Sciences*, 107(8).

Meyerson LA, Saltonstall K, et al. 2000. A comparison of *Phragmites australis* in freshwater and brackish marsh environments in North America. *Wetlands Ecology and Management*, 8.

Ryan ME, Johnson JR & Fitzpatrick BM. 2009. Invasive hybrid tiger salamander genotypes impact native amphibians. *Proceedings of the National Academy of Sciences*, 106(27).

Saltonstall K. 2002. Cryptic invasion by a non-native genotype of the common reed, *Phragmites australis*, into North America. *Proceedings of the National Academy of Sciences*, 99(4).

Vasquez EA, Glenn EP, et al. 2005. Salt tolerance underlies the cryptic invasion of North American salt marshes by an introduced haplotype of the common reed *Phragmites australis* (Poaceae). *Marine Ecology Progress Series*, 298.

Biogeochemistry, Nutrients, and Productivity

Batzer DP & Sharitz RR, eds. 2014. *Ecology of Freshwater and Estuarine Wetlands*. University of California Press.

Keddy PA. 2000. *Wetland Ecology: Principles and Conservation*. Cambridge University Press.

Schlesinger WH & Bernhardt ES. 2013. *Biogeochemistry: An Analysis of Global Change*. Academic Press.

Waide RB, Willig MR, et al. 1999. The relationship between productivity and species richness. *Annual Review of Ecology and Systematics*, 30(1).

General Salt Marsh Ecology
(Zonation, Tolerance, Competition, Phragmites*)*

Bertness MD. 1991. Interspecific interactions among high marsh perennials in a New England salt marsh. *Ecology,* 72(1).

Bertness MD & Shumway SW. 1993. Competition and facilitation in marsh plants. *The American Naturalist,* 142(4).

Dreyer GD & Caplis M, eds. 2001. *Living Resources and Habitats of the Lower Connecticut River.* Bulletin Number 37. The Connecticut College Arboretum.

Emery NC, Ewanchuk PJ & Bertness MD. 2001. Competition and salt-marsh plant zonation: Stress tolerators may be dominant competitors. *Ecology,* 82(9).

Ewanchuk PJ & Bertness MD. 2003. Recovery of a northern New England salt marsh plant community from winter icing. *Oecologia,* 136(4).

Ewanchuk PJ & Bertness MD. 2004. The role of waterlogging in maintaining forb pannes in northern New England salt marshes. *Ecology,* 85(6).

Ewanchuk PJ & Bertness MD. 2004. Structure and organization of a northern New England salt marsh plant community. *Journal of Ecology,* 92(1).

Levine JM, Brewer JS & Bertness MD. 1998. Nutrients, competition and plant zonation in a New England salt marsh. *Journal of Ecology,* 86(2).

Miller WR & Egler FE. 1950. Vegetation of the Wequetequock-Pawcatuck tidal-marshes, Connecticut. *Ecological Monographs,* 20(2).

Odum WE. 1988. Comparative ecology of tidal freshwater and salt marshes. *Annual Review of Ecology and Systematics,* 19(1).

Silliman BR, Grosholz T & Bertness MD, eds. 2009. *Human Impacts on Salt Marshes: A Global Perspective.* University of California Press.

Shumway SW. 1995. Physiological integration among clonal ra-
mets during invasion of disturbance patches in a New England salt
marsh. *Annals of Botany*, 76(3).

Ditching Marshes

Bourn WS & Cottam C. 1951. *Some Biological Effects of Ditching
Tidewater Marshes*. Research Report 19. US Department of the Interior, Fish and Wildlife Service.
Seabold KR. 1992. *From Marsh to Farm: The Landscape Transfor-
mation of Coastal New Jersey*. US Department of the Interior, National Park Service.

Monarch Migration

Howard E & Davis AK. 2009. The fall migration flyways of monarch butterflies in eastern North America revealed by citizen scientists. *Journal of Insect Conservation*, 13(3).

Mangroves

Stuart SA, Choat B, et al. 2007. The role of freezing in setting
the latitudinal limits of mangrove forests. *New Phytologist*, 173(3).

Sea Level Rise and Marsh Development

Donnelly JP & Bertness MD. 2001. Rapid shoreward encroach-
ment of salt marsh cordgrass in response to accelerated sea-level
rise. *Proceedings of the National Academy of Sciences*, 98(25).

Gornitz V. 1995. Sea-level rise: A review of recent past and near-future trends. *Earth Surface Processes and Landforms*, 20(1).

Kirwan ML, Temmerman S, et al. 2016. Overestimation of marsh vulnerability to sea level rise. *Nature Climate Change*, 6(3).

Orson RA, Warren RS & Niering WA. 1998. Interpreting sea level rise and rates of vertical marsh accretion in a southern New England tidal salt marsh. *Estuarine, Coastal and Shelf Science*, 47(4).

Raposa KB, Weber RL, et al. 2017. Vegetation dynamics in Rhode Island salt marshes during a period of accelerating sea level rise and extreme sea level events. *Estuaries and Coasts*, 40(3).

Redfield AC. 1972. Development of a New England salt marsh. *Ecological Monographs*, 42(2).

Schepers L, Kirwan ML, et al. 2020. Evaluating indicators of marsh vulnerability to sea level rise along a historical marsh loss gradient. *Earth Surface Processes and Landforms*, 45(9).

Warren RS & Niering WA. 1993. Vegetation change on a northeast tidal marsh: Interaction of sea-level rise and marsh accretion. *Ecology*, 74(1).

Watson EB, Wigand C, et al. 2017. Wetland loss patterns and inundation-productivity relationships prognosticate widespread salt marsh loss for southern New England. *Estuaries and Coasts*, 40(3).

Leaf Miners

Dorchin N, Joy JB, et al. 2015. Taxonomy and phylogeny of the *Asphondylia* species (Diptera: Cecidomyiidae) of North American goldenrods: Challenging morphology, complex host associations, and cryptic speciation. *Zoological Journal of the Linnean Society*, 174(2).

Salamanders

Bogart JP, Bi K, et al. 2007. Unisexual salamanders (genus *Ambystoma*) present a new reproductive mode for eukaryotes. *Genome*, 50(2).

Charney ND. 2012. Relating hybrid advantage and genome replacement in unisexual salamanders. *Evolution*, 66(5).

Neill WT. 1958. The occurrence of amphibians and reptiles in saltwater areas, and a bibliography. *Bulletin of Marine Science*, 8(1).

7. ELEVATION

Source to Sea

Tougias MJ. 2008. *River Days: Exploring the Connecticut River from Source to Sea.* On Cape Publications.

Mount Cardigan

Baldwin HI. 1974. The flora of Mount Monadnock, New Hampshire. *Rhodora*, 76(806).

Bowman PJ. 2007. *Ecological Inventory of Cardigan Mountain State Forest.* New Hampshire Natural Heritage Bureau DRED—Division of Forests & Lands and The Nature Conservancy.

Forbes CB. 1953. Barren mountain tops in Maine and New Hampshire. *Appalachia*, 19.

Murdock H. 1881. Mount Cardigan. *Appalachia*, 2.

Steele FL & Hodgdon AR. 1973. Two interesting plants on Mt. Cardigan, Orange, New Hampshire. *Rhodora*, 75(801).

Whitney GG & Moeller RE. 1982. An analysis of the vegetation of Mt. Cardigan, New Hampshire: A rocky, subalpine New England summit. *Bulletin of the Torrey Botanical Club*, 109(2).

Elevation Zones and Alpine Plants

Cogbill CV & White PS. 1991. The latitude-elevation relationship for spruce-fir forest and treeline along the Appalachian mountain chain. *Vegetatio*, 94(2).

Jones MT & Willey LL, eds. 2018. *Eastern Alpine Guide: Natural History and Conservation of Mountain Tundra East of the Rockies*. University Press of New England.

Siccama TG. 1974. Vegetation, soil, and climate on the Green Mountains of Vermont. *Ecological Monographs*, 44(3).

Slack NG & Bell AW. 1995. *AMC Field Guide to the New England Alpine Summits*. Appalachian Mountain Club.

Sperduto DD & Nichols WF. 2011. *Natural Communities of New Hampshire. Second Edition*. New Hampshire Natural Heritage Bureau, UNH Cooperative Extension.

Soils and Nutrients

Arnalds A. 1987. Ecosystem disturbance in Iceland. *Arctic and Alpine Research*, 19(4).

DeAngelis KM, Pold G, et al. 2015. Long-term forest soil warming alters microbial communities in temperate forest soils. *Frontiers in Microbiology*, 6.

Gough L, Wookey PA & Shaver GR. 2002. Dry heath arctic tundra responses to long-term nutrient and light manipulation. *Arctic, Antarctic, and Alpine Research*, 34(2).

Haag RW. 1974. Nutrient limitations to plant production in two tundra communities. *Canadian Journal of Botany*, 52(1).

Jacoby R, Peukert M, et al. 2017. The role of soil microorganisms in plant mineral nutrition—current knowledge and future directions. *Frontiers in Plant Science*, 8.

Jonasson S, Michelsen A, et al. 1999. Responses in microbes

and plants to changed temperature, nutrient, and light regimes in the arctic. *Ecology*, 80(6).

Melillo JM, Frey SD, et al. 2017. Long-term pattern and magnitude of soil carbon feedback to the climate system in a warming world. *Science*, 358(6359).

Glacial Striations

Thompson W. 1999. History of research on glaciation in the White Mountains, New Hampshire (USA). *Géographie physique et Quaternaire*, 53(1).

Galápagos

Gibbs JP, L Cayot & W Tapia A, eds. 2021. *Galápagos Giant Tortoises*. Academic Press

8. DISTURBANCE
Geology

Hubert JF. 2017. *Triassic-Jurassic of Western Massachusetts: Easily Accessible Geology Field Trips*. CreateSpace Independent Publishing Platform.

Navigation

Frost BJ & Mouritsen H. 2006. The neural mechanisms of long distance animal navigation. *Current Opinion in Neurobiology*, 16(4).

Phillips JB, Adler K & Borland SC. 1995. True navigation by an amphibian. *Animal Behaviour*, 50(3).

Pittman SE, Hart KM, et al. 2014. Homing of invasive Burmese pythons in South Florida: Evidence for map and compass senses in snakes. *Biology Letters*, 10(3).

Sinsch U. 2006. Orientation and navigation in Amphibia. *Marine and Freshwater Behaviour and Physiology*, 39(1).

Porcupines

Coltrane JA. 2012. Redefining the North American porcupine (*Erethizon dorsatum*) as a facultative specialist herbivore. *Northwestern Naturalist*, 93(3).

Diner B, Berteaux D, et al. 2009. Behavioral archives link the chemistry and clonal structure of trembling aspen to the food choice of North American porcupine. *Oecologia*, 160(4).

Felicetti LA, Shipley LA, et al. 2000. Digestibility, nitrogen excretion, and mean retention time by North American porcupines (*Erethizon dorsatum*) consuming natural forages. *Physiological and Biochemical Zoology*, 73(6).

Morin P, Berteaux D & Klvana I. 2005. Hierarchical habitat selection by North American porcupines in southern boreal forest. *Canadian Journal of Zoology*, 83(10).

Roze U. 2009. *The North American Porcupine*. Cornell University Press.

Schmidt KN & Christian DP. 1988. *Porcupine-Eastern Hemlock Interactions in Hemlock Ravine Scientific and Natural Area*. Final Report to the Nongame Wildlife Program, Minnesota Department of Natural Resources. University of Minnesota-Duluth.

Seigler D & Price PW. 1976. Secondary compounds in plants: Primary functions. *The American Naturalist*, 110(971).

Shapiro J. 1949. Ecological and life history notes on the porcupine in the Adirondacks. *Journal of Mammalogy*, 30(3).

Snyder MA & Linhart YB. 1997. Porcupine feeding patterns:

Selectivity by a generalist herbivore? *Canadian Journal of Zoology*, 75(12).

Hemlock, Woolly Adelgid, and Elongate Scale

Barden LS. 1979. Tree replacement in small canopy gaps of a Tsuga canadensis forest in the Southern Appalachians, Tennessee. *Oecologia*, 44(1).

Ellison AM, Bank MS, et al. 2005. Loss of foundation species: Consequences for the structure and dynamics of forested ecosystems. *Frontiers in Ecology and the Environment*, 3(9).

Havill NP, Vieira LC & Salom SM. 2014. Biology and control of hemlock woolly adelgid. US Department of Agriculture, Forest Service, Forest Health Technology Enterprise Team.

Miller-Pierce MR, Orwig DA & Preisser E. 2010. Effects of hemlock woolly adelgid and elongate hemlock scale on eastern hemlock growth and foliar chemistry. *Environmental Entomology*, 39(2).

Paradis A, Elkinton J, et al. 2008. Role of winter temperature and climate change on the survival and future range expansion of the hemlock woolly adelgid (*Adelges tsugae*) in eastern North America. *Mitigation and Adaptation Strategies for Global Change*, 13.

Record S, McCabe T, et al. 2018. Identifying foundation species in North American forests using long-term data on ant assemblage structure. *Ecosphere*, 9(3).

Deer and Trophic Interactions

DeStefano S, Faison EK, et al. 2010. Forest exclosures: An experimental approach to understanding browsing by moose and deer. *Massachusetts Wildlife*, 60.

Elkinton JS, Healy WM, et al. 1996. Interactions among gypsy moths, white-footed mice, and acorns. *Ecology*, 77(8).

Jones CG, Ostfeld RS, et al. 1998. Chain reactions linking acorns to gypsy moth outbreaks and Lyme disease risk. *Science*, 279(5353).

Levi T, Kilpatrick AM, et al. 2012. Deer, predators, and the emergence of Lyme disease. *Proceedings of the National Academy of Sciences*, 109(27).

McShea WJ. 2000. The influence of acorn crops on annual variation in rodent and bird populations. *Ecology*, 81(1).

Ostfeld RS, Jones CG & Wolff JO. 1996. Of mice and mast. *Bio-Science*, 46(5).

Hurricanes, Tornadoes, and Ice Storms

Boose ER, Chamberlin KE & Foster DR. 2001. Landscape and regional impacts of hurricanes in New England. *Ecological Monographs*, 71(1).

Cannon JB, Hepinstall-Cymerman J, et al. 2016. Landscape-scale characteristics of forest tornado damage in mountainous terrain. *Landscape Ecology*, 31(9).

Cooper-Ellis S, Foster DR, et al. 1999. Forest response to catastrophic wind: Results from an experimental hurricane. *Ecology*, 80(8).

D'Amato AW, Orwig DA, et al. 2017. Long-term structural and biomass dynamics of virgin Tsuga canadensis–Pinus strobus forests after hurricane disturbance. *Ecology*, 98(3).

Everham EM & Brokaw NV. 1996. Forest damage and recovery from catastrophic wind. *The Botanical Review*, 62(2).

O'Connor P. 2006, July 12. Was it a tornado? *The Greenfield Recorder*.

Plotkin AB, Foster D, et al. 2013. Survivors, not invaders, control forest development following simulated hurricane. *Ecology*, 94(2).

Rustad LE & Campbell JL. 2012. A novel ice storm manipulation experiment in a northern hardwood forest. *Canadian Journal of Forest Research*, 42(10).

Ecology of Fallen Trees

Brang P, Moran J, et al. 2003. Regeneration of *Picea engelmannii* and *Abies lasiocarpa* in high-elevation forests of south-central British Columbia depends on nurse logs. *The Forestry Chronicle*, 79(2).

Idol TW, Figler RA, et al. 2001. Characterization of coarse woody debris across a 100 year chronosequence of upland oak–hickory forests. *Forest Ecology and Management*, 149.

MacMillan PC. 1988. Decomposition of coarse woody debris in an old-growth Indiana forest. *Canadian Journal of Forest Research*, 18(11).

Plotkin AB, Schoonmaker P, et al. 2017. Microtopography and ecology of pit-mound structures in second-growth versus old-growth forests. *Forest Ecology and Management*, 404.

Russell, MB, Woodall CW, et al. 2014. Residence times and decay rates of downed woody debris biomass/carbon in eastern US forests. *Ecosystems*, 17(5).

Old Growth, Management, and History

Caputo J & D'Amato T. 2006. *Mount Toby Demonstration Forest Management Plan*. Department of Forestry and Wildlife Management, University of Massachusetts Amherst.

D'Amato AW, Orwig DA & Foster DR. 2006. New estimates of Massachusetts old-growth forests: Useful data for regional conservation and forest reserve planning. *Northeastern Naturalist*, 13(4).

D'Amato AW, Raymond P & Fraver S. 2018. Old-growth disturbance dynamics and associated ecological silviculture for forests in northeastern North America. In Barton AM & Keeton WS., eds., *Ecology and Recovery of Eastern Old-Growth Forests* (pp. 99–118). Island Press.

Keeton WS, Lorimer CG, et al. 2018. Silviculture for eastern old growth in the context of global change. In Barton AM & Keeton WS., eds., *Ecology and Recovery of Eastern Old-Growth Forests* (pp. 237–265). Island Press.

Lyons LM. 1923, October 7. Mt. Tobey, found again, to celebrate Friday. *Boston Globe.*

Smith JM. 1899. *History of the Town of Sunderland, Massachusetts.* E. A. Hall.

Wilson BF. 1987. *The Birth of Two University Forests.* Department of Forestry and Wildlife Management, University of Massachusetts Amherst.

Wilson BF. 2017. *Happy Birthday Mount Toby Demonstration Forest 1916–2016. Why You See What You See.* Sunderland Public Library.

Wildlife Dynamics over Time

Foster DR, Motzkin G, et al. 2002. Wildlife dynamics in the changing New England landscape. *Journal of Biogeography,* 29.

9. RELICS
Cattails

Mitich, LM. 2000. Common cattail, *Typha latifolia* L. *Weed Technology,* 14(2).

Pitcher Plants

Butler JL, Atwater DZ & Ellison AM. 2005. Red-spotted newts: An unusual nutrient source for northern pitcher plants. *Northeastern Naturalist,* 12(1).

Ellison AM & Gotelli NJ. 2002. Nitrogen availability alters the expression of carnivory in the northern pitcher plant, *Sarracenia purpurea*. *Proceedings of the National Academy of Sciences*, 99(7).

Sphagnum Chemistry

Andrus RE. 1986. Some aspects of sphagnum ecology. *Canadian Journal of Botany*, 64(2).

Hemond HF. 1980. Biogeochemistry of Thoreau's bog, Concord, Massachusetts. *Ecological Monographs*, 50(4).

McLellan JK & Rock CA. 1988. Pretreating landfill leachate with peat to remove metals. *Water, Air, and Soil Pollution*, 37.

Painter TJ. 1991. Lindow man, Tollund man and other peat-bog bodies: The preservative and antimicrobial action of Sphagnan, a reactive glycuronoglycan with tanning and sequestering properties. *Carbohydrate Polymers*, 15(2).

Urban NR & Bayley SE. 1986. The acid-base balance of peatlands: A short-term perspective. *Water, Air, and Soil Pollution*, 30.

Chestnuts

Foster DR & Zebryk TM. 1993. Long-term vegetation dynamics and disturbance history of a Tsuga-dominated forest in New England. *Ecology*, 74(4).

Kittredge J. 1913. Notes on the chestnut bark disease in Petersham, Massachusetts. *Harvard Forestry Club Bulletin*, 2.

Wetland Development and Hydrology

Anderson RL, Foster DR & Motzkin G. 2003. Integrating lateral expansion into models of peatland development in temperate New England. *Journal of Ecology*, 91(1).

Brooks RT. 2005. A review of basin morphology and pool hydrology of isolated ponded wetlands: Implications for seasonal forest pools of the northeastern United States. *Wetlands Ecology and Management*, 13(3).

Brooks RT. 2009. Potential impacts of global climate change on the hydrology and ecology of ephemeral freshwater systems of the forests of the northeastern United States. *Climatic Change*, 95.

Gaudig G, Couwenberg J & Joosten H. 2006. Peat accumulation in kettle holes: Bottom up or top down. *Mires and Peat*, 1(6).

Wieder RK, Novák M, et al. 1994. Rates of peat accumulation over the past 200 years in five sphagnum-dominated peatlands in the United States. *Journal of Paleolimnology*, 12(1).

Wilcox DA & Simonin HA. 1988. The stratigraphy and development of a floating peatland, Pinhook Bog, Indiana. *Wetlands*, 8(1).

Island Biogeography and Metapopulation Theory

Charney ND. 2012. Evaluating expert opinion and spatial scale in an amphibian model. *Ecological Modelling*, 242.

Hanski I. 1998. Metapopulation dynamics. *Nature*, 396(6706).

MacArthur RH & Wilson EO. 1963. An equilibrium theory of insular zoogeography. *Evolution*, 17(4).

Marsh DM & Trenham PC. 2001. Metapopulation dynamics and amphibian conservation. *Conservation Biology*, 15(1).

Simberloff D & Abele LG. 1982. Refuge design and island biogeographic theory: Effects of fragmentation. *The American Naturalist*, 120(1).

Chytrid Fungus

Cheng TL, Rovito SM, et al. 2011. Coincident mass extirpation of neotropical amphibians with the emergence of the infectious fungal pathogen *Batrachochytrium dendrobatidis*. *Proceedings of the National Academy of Sciences*, 108(23).

Martel A, Blooi M, et al. 2014. Recent introduction of a chytrid fungus endangers Western Palearctic salamanders. *Science*, 346(6209).

Martel A, Spitzen-van der Sluijs A, et al. 2013. *Batrachochytrium salamandrivorans* sp. nov. causes lethal chytridiomycosis in amphibians. *Proceedings of the National Academy of Sciences*, 110(38).

Mendelson III JR, Jones ME, et al. 2014. On the timing of an epidemic of amphibian chytridiomycosis in the highlands of Guatemala. *South American Journal of Herpetology*, 9(2).

Rovito SM, Parra-Olea G, et al. 2009. Dramatic declines in neotropical salamander populations are an important part of the global amphibian crisis. *Proceedings of the National Academy of Sciences*, 106(9).

Spitzen-van der Sluijs A, Stegen G, et al. 2018. Post-epizootic salamander persistence in a disease-free refugium suggests poor dispersal ability of *Batrachochytrium salamandrivorans*. *Scientific Reports*, 8(1).

New Madrid Earthquakes

Penick JL. 1981. *The New Madrid Earthquakes*. University of Missouri Press.

10. ALIENS
Invasive Species and New Zealand Fauna

Alpert P. 2006. The advantages and disadvantages of being introduced. *Biological Invasions*, 8(7).

Cipollini D & Cipollini K. 2016. A review of garlic mustard (*Al-*

liaria petiolata, Brassicaceae) as an allelopathic plant. *Journal of the Torrey Botanical Society*, 143(4).

Clout M. 2001. Where protection is not enough: Active conservation in New Zealand. *Trends in Ecology & Evolution*, 16(8).

Department of Conservation. 2011. *Trounson Kauri Park Mainland Island Annual Report 2010/11*. Dargaville, New Zealand.

Eggleston JE, Rixecker SS & Hickling GJ. 2003. The role of ethics in the management of New Zealand's wild mammals. *New Zealand Journal of Zoology*, 30(4).

Mark AF. 1985. The botanical component of conservation in New Zealand. *New Zealand Journal of Botany*, 23(4).

Narango DL, Tallamy DW & Marra PP. 2018. Nonnative plants reduce population growth of an insectivorous bird. *Proceedings of the National Academy of Sciences*, 115(45).

Sax DF, Gaines SD & Brown JH. 2002. Species invasions exceed extinctions on islands worldwide: A comparative study of plants and birds. *The American Naturalist*, 160(6).

Snow N & Witmer G. 2010. *American bullfrogs as invasive species: A review of the introduction, subsequent problems, management options, and future directions*. Staff Publications. US Department of Agriculture, Wildlife Research Center.

Stachowicz JJ, Whitlatch RB & Osman RW. 1999. Species diversity and invasion resistance in a marine ecosystem. *Science*, 286(5444).

Whale Rescue

Ogle M. 2017. Managing the welfare of marine mammals at mass strandings in Golden Bay, New Zealand. In Butterworth A, ed., *Marine Mammal Welfare: Human Induced Changes in the Marine Environment and Its Impacts on Marine Mammal Welfare* (pp. 137–146). Springer.

Salamander Road Crossings

Gibbs JP & Shriver WG. 2005. Can road mortality limit populations of pool-breeding amphibians? *Wetlands Ecology and Management*, 13(3).

Jackson SD & Tyning TF. 1989. Effectiveness of drift fences and tunnels for moving spotted salamanders *Ambystoma maculatum* under roads. In Langton TES, ed., *Amphibians and Roads, Proceedings of the Toad Tunnel Conference*. ACO Polymer Products, Shefford, England.

Red Foxes

Kamler JF & Ballard WB. 2002. A review of native and nonnative red foxes in North America. *Wildlife Society Bulletin*, 30(2).

Statham MJ, Sacks BN, et al. 2012. The origin of recently established red fox populations in the United States: Translocations or natural range expansions? *Journal of Mammalogy*, 93(1).

Voles

Bergeron JM & Jodoin L. 1987. Defining "high quality" food resources of herbivores: The case for meadow voles (*Microtus pennsylvanicus*). *Oecologia*, 71(4).

Gilbert S, Norrdahl K, et al. 2013. Vole damage to woody plants reflects cumulative rather than peak herbivory pressure. *Annales Zoologici Fennici* 50(4).

Gill RMA. 1992. A review of damage by mammals in north temperate forests: 3. Impact on trees and forests. *Forestry*, 65(4).

Hansson L. 1991. Bark consumption by voles in relation to mineral contents. *Journal of Chemical Ecology*, 17(4).

Krojerová-Prokešová J, Homolka M, et al. 2018. Patterns of vole gnawing on saplings in managed clearings in Central European forests. *Forest Ecology and Management*, 408.

Manson RH, Ostfeld RS & Canham CD. 2001. Long-term effects of rodent herbivores on tree invasion dynamics along forest–field edges. *Ecology*, 82(12).

Ostfeld RS, Manson RH & Canham CD. 1997. Effects of rodents on survival of tree seeds and seedlings invading old fields. *Ecology*, 78(5).

Palo RT. 1984. Distribution of birch (*Betula* spp.), willow (*Salix* spp.), and poplar (*Populus* spp.) secondary metabolites and their potential role as chemical defense against herbivores. *Journal of Chemical Ecology*, 10(3).

Pigott CD. 1985. Selective damage to tree-seedlings by bank voles (*Clethrionomys glareolus*). *Oecologia*, 67(3).

Schreiber LA & Swihart RK. 2009. Selective feeding of pine voles on roots of tree seedlings. *Canadian Journal of Zoology*, 87(2).

Rose Hips and Itchy Butts

Chrubasik C, Roufogalis BD, et al. 2008. A systematic review on the Rosa canina effect and efficacy profiles. *Phytotherapy Research*, 22(6).

Venkatesh RP, Ramaesh K & Browne B. 2005. Rose-hip keratitis. *Eye*, 19(5).

Robin Migration

Brown D & Miller G. 2016. Band recoveries reveal alternative migration strategies in American robins. *Animal Migration*, 3(1).

Lafleur NE, Rubega MA & Elphick CS. 2007. Invasive fruits,

novel foods, and choice: An investigation of European starling and American robin frugivory. *The Wilson Journal of Ornithology*, 119(3).

Lundberg P. 1988. The evolution of partial migration in birds. *Trends in Ecology & Evolution*, 3(7).

Suthers HB, Bickal JM & Rodewald PG. 2000. Use of successional habitat and fruit resources by songbirds during autumn migration in central New Jersey. *The Wilson Bulletin*, 112(2).

Bird Language

Searcy WA & Nowicki S. 2005. *The Evolution of Animal Communication: Reliability and Deception in Signaling Systems.* Princeton University Press.

Young J. 2012. *What the Robin Knows: How Birds Reveal the Secrets of the Natural World*. Houghton Mifflin Harcourt.

New England Cottontail

Brown AL & Litvaitis JA. 1995. Habitat features associated with predation of New England cottontails: What scale is appropriate? *Canadian Journal of Zoology*, 73(6).

Dalke PD & Sime PR. 1941. Food habits of the eastern and New England cottontails. *The Journal of Wildlife Management*, 5(2).

Litvaitis JA, Barbour MS, et al. 2008. Testing multiple hypotheses to identify causes of the decline of a lagomorph species: The New England cottontail as a case study. In Alves PC, Ferrand N & Hackländer K, eds., *Lagomorph Biology: Evolution, Ecology, and Conservation* (pp. 167–185). Springer.

Litvaitis JA, Tash JP, et al. 2006. A range-wide survey to determine the current distribution of New England cottontails. *Wildlife Society Bulletin*, 34(4).

Coyotes

Hody JW & Kays R. 2018. Mapping the expansion of coyotes (*Canis latrans*) across North and Central America. *ZooKeys*, 759.

Way JG, Rutledge L, et al. 2010. Genetic characterization of eastern "coyotes" in eastern Massachusetts. *Northeastern Naturalist*, 17(2).

Bobcats

Litvaitis JA, Sherburne JA & Bissonette JA. 1986. Bobcat habitat use and home range size in relation to prey density. *The Journal of Wildlife Management*, 50(1).

Ryden H. 1981. *Bobcat Year.* Viking Press.

Opossum

Kanda LL, Fuller TK, et al. 2009. Seasonal source–sink dynamics at the edge of a species' range. *Ecology*, 90(6).

Eastern Cougars

Allardyce G. 2001. *On the Track of the New Brunswick Panther: The Story of Bruce Wright and the Eastern Panther.* Self-published.

Bolgiano C & Roberts J, eds. 2005. *The Eastern Cougar: Historic Accounts, Scientific Investigations, and New Evidence.* Stackpole Books.

Hawley JE, Rego PW, et al. 2016. Long-distance dispersal of a subadult male cougar from South Dakota to Connecticut documented with DNA evidence. *Journal of Mammalogy,* 97(5).

Index

Page references in *italics* refer to illustrations.

ing habits of, 120; gait of, 335; marking behavior of, 106

eastern hemlock, ix; and elevation, 210; and elongate hemlock scale, *253*, 253–54; observing for disturbance to, 241–44, *244*, 247–49, 267–71; removal experiments, *255*, 255–58; in wetlands ("Hemlock Hollow"), 285; and woolly adelgid, 35, *36*, 40, 251–54, *252*

eastern red-spotted newt, *159*

ecology specialties: paleoecologists, 161, 284; plant ecologists, 3–7, 10–11, 31–32. *See also* biogeography; conservation biology; *Field Naturalist* (college course); Harvard Forest; tracking

edge habitat, 120, 134–41

egret, *166*

Einstein, Albert, 182–85

Eiseman, Charley: on bird alarm calls, 327; fens researched by, 286–88; salt marsh research of, 195–97; tracking by, 10–11, 15, 19, 76–77, 106, 119–20, 185–86; tracking by, and conservation biology, 316–26, *322*, *324*, *325*, 337–38; at White Sands National Monument, 254

elevation, 204–37; Connecticut River's alpine summit, overview, 207–8; elevation gradient and temperature contrast of mountainous regions, 220–26, *222–27*, 228–34, *231*; and glacial till, 111–13; and Mount Cardigan characteristics, 208, 214–17, *216*, *217*, 220–21, 230–32, 234–37, *235*, *236*; and Mount Washington

characteristics, 207–8, 220–26, *222–25*, 228–32; and transition from boreal forests to broadleaf forests, *204–5*, 214–18, 221–26, *222*, *223*; and tundra climate, *225*, 226–28, 232

elm tree, 324, *324*

elongate hemlock scale, *253*, 253–54

endangered species: and climate change, 126, 132–34, *133*; versus endangered ecosystems, 138–39; and land clearing, 262; laws protecting, 139; Puritan tiger beetle, 124, 128–35, *129*, *133*, 138

epicormic sprouting, *66*, 67

evaporation, 28–29

evolution, 173, 201–3

fairy shrimp, *99*

Faison, Ed, 41, 42

farming: Maggie's Forest history, 92; salt hay, 168, 170, 175–76, 189–94, 198–200. *See also* The Orchard

feldspar, 45, 50, 54

fens: defined, 286–87; as diverse habitat, 287–89, *288*; former spruce forest example, *272–73*, 286–87, 299

ferns: bracken, ix; royal, ix, 102

Fiedler, Peggy, 31–32

Field Naturalist (college course): on conservation biology, 305, 352; on disturbance to forests, *244*, *248*, 248–49; on elevation, *226*; on orientation skills, 148; overview, 18–21, *21*; on recognizing left-right pattern in topography, 26–27; rocks identified in, 52;

200, *196*, *199*; Virginia glasswort, ix
Gleason, Henry, 78
glowworm beetle, 212
GMO (genetically modified organisms), 43
Gondwana, 158–63
GPS devices, 101, 147–49
granite, ix; and topography, 45, 50,
51, 54
gravity and tidal force, 182–85
gray fox, 284, *343*
gray squirrel: behavior of, 105; gait
of, 318–19; tree marking by, 15
great egret, 165, *166*
greater yellowlegs, 164, *164*
Great New England Hurricane of
1938, *258*, 259
Great Swamp: elevation of, 111–13;
forest contrast in, 102–4; water
and clay layers in, 279
groundhog, 105
Guanwu Formosan salamanders,
159, *159*
Gwiazdowski, Rodger, 124, 127

habitat fragmentation, 145–46, 294
Harms, Tekla, 143
Harvard Forest: as author's work
base, 233; and Foster, 284; hurricane simulation in, *258–61*,
258–67; pitcher plant research,
280; soil-warming experiment,
233–34
Harvester, Hannah, 328
Haudenosaunee five nations, 1,
20, 65
hemlock. *See* eastern hemlock
hepatica. *See* round-lobed hepatica
highbush blueberry, ix; in fens,
287; and spatial context, 97

hog peanut, ix, *30*
home environment, lawns, and
landscaping, x–xii, 1–21; barren yards, *9*, 12; effect on conservation, 4–5, 80–81, 345–47,
351–53; interpreting landscape
of home environment, 16–21,
21; lawns versus desert habitat, 4–12, *8–11*; mesic yards,
6, *9*, 12; natural environment
and socioeconomic status, 12–
14; trees and benefit to yards,
xii, 1–2; urban settings and
invasive-dominated degraded
eco-systems, 14–16; wild mesic yards, 12; wild xeric yards,
8; xeric yards, 6, *8*, 12; versus
yards as habitat, 2–4, *3*. *See also*
disturbance; spatial context
hornet, *288*
house finch, 13
house sparrow, 7, *10*
Hubbard Brook (New Hampshire),
257
humans, ix; artifacts of, moved by
topography, 135–38, 151; population statistics, 17. *See also*
home environment, lawns, and
landscaping
hunting of invasive species, 308–9
hurricanes, *258–61*, *258–67*, *264–66*

Iceland, climate of, 226–28
ice storm research, 257
igneous rocks, 52–53
Indiana Dunes National Lakeshore, *69*, *76*, 76–80
Indian cucumber, 214
insects: beech scale, 108; behavior and topography change,
124, 127–35, *129*, *131*, *133*, 138;

mal behavior, 117–21, *119*; and cut-banks in meandering rivers, 148–51; and displaced fish species, 124–27; endangered species versus endangered ecosystems, 138–39; human artifacts moved by, 135–38, 151; and insect behavior, 124, 127–35, *129*, *131*, *133*, 138; maps as evidence of, 121–23, *122*, 149–51, *150*; and oxbow lakes, *122*, 149; and riparian forests, 139–41, 144–48; rivers created by glacial lakes, 117, *118*; and terraces created by rivers, 141–44

mesic yards, 6, *9*, 12

metamorphic rocks, 52–53, 157

metapopulation theory, 292, 294

mice, 257, 334–37, *336*

microbursts, 268

microclimate, 29

migraine, ocular, 350

Milky Way, 28

minerals, and plate tectonics, 46. *See also names of individual minerals*

monarch butterfly, 170

moose, ix; beds made by, 256; feeding habits, 324; in fens, 287; marking behavior of, 105

Morse, Sue, 107

mosquito, 100, 173, 175–76

moths: and biogeography, 186–88, 195–97, *196*; case-bearing moth, *196*, 197; pygmy leaf mining moth, ix, 196–97; spongy moth, 245, 257, 269

mountain laurel, ix; habitat preference, *30*; and topography, 41

mountain lion (eastern cougar), 348–51

Mount Cardigan, 208, 214–17, *216*, *217*, 220–21, 230–32, 234–37, *235*, 236, *236*. *See also* elevation

Mount Washington, 207–8, 220–26, *222–25*, 228–32. *See also* elevation

mowing. *See* home environment, lawns, and landscaping

mullein, 208–10

Müller, Paul Hermann, 172–73

multiflora rose, ix; as invasive species, 306–7, 334

multiple-trunk trees, 15, 32–33, *33*, 85, 92, 269

muskrat, 278

Native Americans: Algonquian languages and wigwam, 186; arrowheads made by, 50–51; Calusa Indians, 136–37; glacial lake lore of, 89; Haudenosaunee five nations, 1, 20, 65; landscape practices of, 81–82

natural communities, 29

nature, valuing, 309, 353–54. *See also* conservation biology

Navajo sandstone, 54

navigation, 25–26, 147–48, 168, 217

New England cottontail, 330–31

New Hampshire, Hubbard Brook research, 257. *See also* Mount Cardigan; Mount Washington

newts, *159*, 159–60, 250, 293

New Zealand: beech trees of, 160–61, *161*; and invasive species, 308–9; whale rescue operation in, 309–14, *310*

nitrogen, 194, 280–81

Norris, John, 104

North America, plate tectonics and formation of, 46

Northern Hemisphere, sun aspect, 28
northern red oak, ix; characteristics of, 102; and forest disturbance, 259, 262, 268–69
nurse logs, 40, 265, *266*
nutrient flows and chemical cycles, 169–81; acidic environment of wetlands, 279–86, *281–83*; chemical defenses of trees, *325*, 325–26; ditching/draining marsh water for agriculture, 175–76, 189–94, 198–200; and insecticides, 172–75; natural migration of nutrients, 169–72, *170*, *172*, 178–81, *179*; nutrient decomposition, 232; nutrients in fens versus bogs, 287; pathogens and effect to ecological systems, 32–42, *33*, *36*, *42*, 47–49; soil nutrients, overview, 53–58, *55*; and tidal force, 177–78, 181–85. *See also* zonation

oak gall wasp, ix
oak trees: bear oak, 68; characteristics of, 68–69; dwarf chestnut oak, 68; and forest succession, 70–80, *76*, *79*; northern red oak, ix, 102, 259, 262, 268–69; versus pine tree pattern, overview, 64; pin oak, ix, 102–4, 340; savannas, 68–69, *69*; scarlet oak, ix, 68; swamp white oak, ix, 102; and wildlife management, 83, 85. *See also* forest succession
old-growth. *See* forest succession
opossum, 338
The Orchard, and invasive plant species as habitat, *302*, 303–9,

304–6, 345–47, 351–53. *See also* conservation biology
oriental bittersweet, ix, *306*, 306–7
orientation. *See* aspect and microclimates; navigation; spatial context
osprey, ix, 172–75
otter, 106
oxbow lakes, *122*, 149
oxygen, 190–91

Pangea, 162
pannes, 198–99, *199*
paper birch, ix, *30*. *See also* birch trees
paper-making wasp, ix, xii
pathogens. *See* chytrid fungus; invasive species
patterns, random versus clumped, 165–66
Patterson, Bill, 245
peat and peatland, 282–87, 295
Pepin, Dan, 245
Peterson, Roger Tory, 172, 174
pH of soil, 57
pigeon, 7, *10*, 14
pine trees: gray squirrels and marking of, 15; herbivore preferences and forest composition, *325*, 326; pine barrens, 68, 83, *85*, 85–86, 90–91; red pine, *74*, *75*, 233. *See also* forest succession; pitch pine; white pine
pin oak, ix; and bobcat marking behavior, 340; characteristics of, 102–4
pitcher plant. *See* purple pitcher plant
pitch pine, ix; and bedrock, 86; characteristics of, 65–70, *66*,

148; fire disturbance and loss of, 235–36; nutrients in, overview, 53–58, *55*; pH of, 57; "rich" soil, 55–58; sediment and glacial till, *90*, 111–13; sediment sorted by water, 87–91, *90*; soil horizons, 56; soil-warming experiment, 233–34. *See also* elevation; nutrient flows and chemical cycles

solar noon, 27. *See also* sun

sorting (geologic), 54, 87–89

southern beech trees, 160–61, *161. See also* American beech trees

Southern Hemisphere, star constellations, 27–28

spatial context, 94–113; and animal behavior, 104–8; and forest contrast/history, 101–4, *109*, 109–10; and glacial till, 110–13; and identifying your location, 25–28, *96*, 97–98; orienting your place in landscape via, 97–98; overreliance on technology for, 147–48; topographic maps for, 100–101; and topography, 25–28, 100–101; of vernal pools, *99*, 99–100

species checklist, ix

sphagnum moss, 281–87, *282*, 299

Spitzer, Paul, 174

spongy moth, 245, 257, 269

spotted salamander, 314–16, *315*

spruce trees: and elevation, 210–11, 214, 221, 223–26, *226*, 230, 237; fen as former spruce forest, *272–73*, 286–87, 299; red spruce and black spruce, 286; white spruce, ix

staghorn sumac, 334

star constellations, Southern Hemisphere, 27–28

stone walls, 110, *111*

striped skunks, ix, 352

Struijk, Richard, 159–60

succession. *See* forest succession

sugar maple, ix, 102

sumac: fragrant, *76*; poison sumac, ix; staghorn, 334

sun: for orientating yourself to location, 25–28; solar noon, 27

sundew, *283*, 286

swamps, 102, 113. *See also* Great Swamp; salt marshes

swamp white oak, ix, 102

Taconic (Bronson Hill) island arc, 157, 242

taiga, 224. *See also* boreal forest biome

Tail of the Dragon (Great Smoky Mountains), 37–38

tawny cottongrass, ix, 286

Tennessee, Reelfoot Lake of, 299

terraces, 141–44

terranes, 157

theory of relativity, 182–85

threadleaf sundew, *283*

Tiburon mariposa lily, 31–32

tides and tidal force, 177–78, 181–85

tidal marsh. *See* salt marshes

tools: adzes, *52*; compasses, 148; knives, 37; overreliance on, 147–48; soil corers, 86

topography, 22–59; Eastern Border fault, 242; geology and plate tectonics, 44–47, 49–53, *52*; glaciers and effect on, 58–59, 298–99; maps and subsequent change to, 100–101, 121–23, *122*,

topography (*continued*)
143, 149–51, *150. See also* maps; meandering rivers
tornados, 245–47, *258, 259*
tracking: animal alarm call recognition, 326, 329; animal bite pattern recognition, 329, 331; animal gait types, 316–26, *322, 324, 325,* 335, 337–38; of insects, 10–11; and marking behavior, 15, 105–6, 284, 321, 337; scat at saddle points, 44; scat clues, 104–6. *See also* conservation biology; *names of individual animals*
trails, location of, 43–44
trees: benefits of, in yards, 1–2; biogeography of, 158, 166; canopy versus understory, 43; chemical defense of, *325,* 325–26; deadwood as habitat, *xii, 288,* 288–89; and evaporation, 29; and forest contrast/history, 101–4, *109,* 109–10; growth from seedlings to maturity, 103–4; identifying age of, 34, 69, 128, 151, 241, 247, 269; nurse logs, 40, 265, *266;* pine versus oak pattern, *62, 64,* 65–70, *66, 67, 69,* 91–92; riparian forests, 139–41, 144–48; scars on, *109,* 109–10, 234–35; snags, 35, *42,* 276, *276,* 288–89; transition from boreal forests to broadleaf forests, *204–5,* 214–18, 221–26, *222, 223;* "tree of peace," 1; "whammy trees," 275; and wind, *224,* 226, 232; "witness trees," 41; wood-sourced heat, 16. *See also* biogeography; conservation bigeography; conservation bi-

ology; disturbance; elevation; forest succession; home environment, lawns, and landscaping; logging; spatial context; topography; water; wetlands
trilobite, 162
trophic levels, 173–74
tulip poplars: characteristics of, *75;* for fire by friction, 210; and topography, 38–39, 40
tundra: alpine, 210, 215, *216,* 218–20, *219, 220,* 225, 230–32, 234–36, *235;* arctic, 218–20, *219,* 226–28, 231, 232; and sedges, 285
tupelo (black gum), ix, 102–3, 106

understory: versus canopy, 43, 73; shrubby characteristics of, *68,* 68–69
unisexual salamanders, *159,* 201–2
United States Geological Survey (USGS), 58, 100–101, 121–23, *122,* 143, 149–51, *150*
urban ecology: Boston example, 14–16; Phoenix example, 4–12, *8–11. See also* home environment, lawns, and landscaping
urine: bobcat, 105–6, 337, 344; fox, 284, 321. *See also* marking behavior; tracking
US Fish and Wildlife Service, 124

valley slopes, left-right pattern of, 26–31, *30,* 43–44
vernal pools, *99,* 99–100, 147–48
Virginia glasswort, ix. *See also* glasswort
volcanic activity, 46–47, 50, 57. *See also* geology
vole, 71, 119–20, 320–26, *322*

wood frog (*continued*)
  fungus, 293; vernal pools and
  reproduction, *99*, 99–100
wood nettle, ix, 145
The Woodsy Club, 20–21, 27, 188,
  303, 320, 331–34
woolly adelgid, 35, 251–54, *252*

xeric yards, 6, *8*, 12

yew, 257
Young, Jon, 208

zonation, 188–201; and "clumped"
  patterns, 165–66; defined, 192–
  93; disturbance to, 195–201, *196*,
  199; and invasive plant species,
  188–89, *189*, 194–95; nutrients
  and effect on, 190–95, *191*; and
  tides, 177–78, 181–85